Guide to Electric Load Management

Guide to Electric Load Management

A. J. Pansini
K. D. Smalling

Tulsa, OK

PennWell Publishing Company
1421 S. Sheridan Road/P.O. Box 1260
Tulsa, OK 74101

Library of Congress Cataloging-in-Publication Data

Pansini, Anthony, and Smalling, Ken
Guide to Electric Load Management / Pansini & Smalling
p. cm.
Includes index.
1. 2.
I. Title

Printed in the United States of America.

1 2 3 4 5 02 01 00 99 98

Dedication

The authors dedicate this book to the men and women of the Long Island Lighting Company who engineered, constructed, operated, and maintained an electric system that experienced a remarkable growth in the pre-Shoreham years. During a relatively short time, from 1945 to 1975, the system peak grew by more than 2,500 megawatts and more than 700,000 new customers were connected.

LILCO people found the ways to meet a challenge rarely experienced by a suburban utility. Innovation, ingenuity, and a devotion to giving the best possible service at the least cost were the trademarks of this workforce. It resulted in many industry firsts, such as triplex secondary, 15-kV load break elbows, plastic-covered primary conductors to reduce tree interruptions, distribution automation controls, common trench installations of electric, gas, and telephone lines, real-time capability of transmission cables and substation transformers, and many more.

The Long Island Lighting Company ended as a corporate entity in June 1998. However, the spirit of those who met the challenge while working together with a feeling of family lives on and will be remembered by those who made the extraordinary achievement possible.

Contents

Preface

This book describes changes in the dynamic electric utility industry—in simple language and with our reasoning for these changes—reflecting the impact of deregulation, environmental regulation, and the competition prevalent in the planning and operation of electric systems.

It joins our companion books—*Guide to Electric Power Distribution Systems, Basics of Electric Power Transmission, Guide to Electric Power Generation, Undergrounding Electric Lines*, and others. To these were added technical works—*Electric Distribution Engineering, High Voltage Equipment Engineering, Power Systems Stability Handbook*, and others.

Initially these were adjuncts to training programs for physical, clerical, and management personnel needed in great numbers to meet the demands of post-World War II America. Updated periodically, the scope of these works expanded beyond their training mission to that of providing technical information essential to those in the government, banking, legal, and academic professions. The scope also expanded to include utility material and equipment suppliers and utility executives. The success of these works is a matter of pride to their authors, who now with similar pride offer this latest work—not only to those mentioned above, but also to consumers, who are becoming directly involved in the electric industry that so completely impacts our lives.

We extend our thanks for assistance and encouragement to our publishers, our former associates, and others who have helped with this work. In particular, we are grateful to our families for their patience during this period and to Marie Vanacore for her secretarial help.

A.J. Pansini
Waco, Texas

K.D. Smalling
Northport, New York

1

The Problem

The electrical utility industry as the basic supplier of electrical energy to consumers has had to cope with significant changes. These changes have occurred in regulatory restrictions, public opinion, and increasing costs—particularly in relation to new generation capacity—from the '70s to the '90s. These changes have created a major problem of matching consumer loads with capacity to supply energy in an economical and reliable manner. The advent of independent power producers, deregulation, and free access to transmission systems owned and operated by utilities have aggravated this problem and added to its complexity.

There has been a dramatic decrease of 40% in net generating capacity additions in this time period, while energy consumption has increased by more than 50%. Without new generating capacity being added to match load growth as in previous years, other means had to be developed to reduce peak demands and maintain an adequate ratio between capacity and demand. An important technology in use is managing consumer loads, and ultimately system loads, by various strategies and techniques made possible by an array of technology advances. This book is intended to provide a general knowledge of demand control and energy conservation generically referred to as electric load management.

Electric Utility Development

The electric utility as the basic supplier of electrical energy is perhaps unique in that almost everyone does business with it and is dependent upon its product. Unlike other enterprises producing commodities, it is obligated to have energy available to meet demands but cannot store up inventory or delay supplying it until it is available.

After the invention of the electric light in the late 1800s, electric energy requirements grew as the innovation caught on. Electricity was used in place of candles and oil lamps as well as replacing mechanically driven machinery with electric drives. Electrically driven refrigerators replaced coolers and iceboxes. Electric elevators resulted in taller and taller buildings. Electric washers, dryers, vacuum cleaners, and cooking ranges changed the lives of homemakers. Electric air conditioners made a significant impact on electric systems in later years, creating summer peaks in most utilities. In the last few decades the electronic age introduced television, computers, and mobile communication. There are few facets of daily living and business today not dependent on a constantly available source of electric energy.

Initially, isolated small electric companies with generators powered by hydro or fossil fired boilers served their consumers over low voltage lines direct from the generator. With the introduction of alternating current (AC), isolated small utilities began to consolidate with others nearby, and the service areas grew larger. The size of generators increased, creating a need for larger central stations and higher voltages to bring this power to consumers.

Steady load growth encouraged by the utilities, some selling appliances, resulted in higher transmission voltage levels. This permitted individual utility areas to interconnect and exchange power for economic reasons as well to provide support under emergency conditions. Load growth accelerated after World War II with many large new base-load generators and many miles of new transmission lines keeping pace with the increase in demands.

Until the 1970s, new generation and transmission facilities were added fairly easily and economically. There was little risk that costs would get out of hand before the project was completed, or that changes in public opinion or environmental regulations would cause significant delays.

As shown in Figure 1–1, U.S. load growth continued to rise—despite the oil embargo and resultant higher energy cost—and apparently will continue to rise in the future, although at a reduced rate.

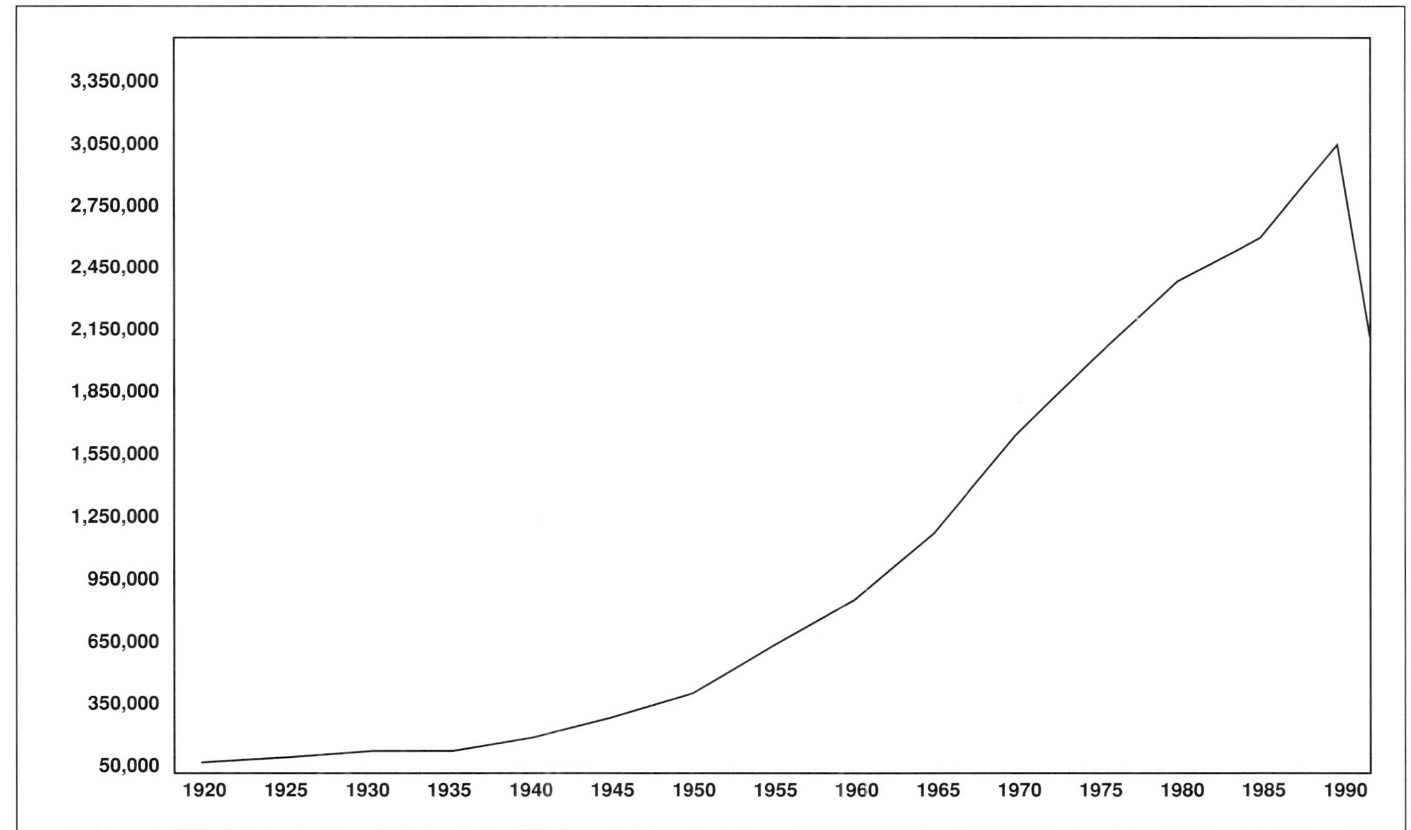

Fig. 1–1 Energy available in the United States, 1920–1990 (millions of kWh)

Capacity vs. Load

It would appear that capacity was keeping pace with load growth during most of this period, but starting in the '70s, the net capacity additions started to decrease as shown in Table 1–1.

In a similar trend, bulk transmission line additions (220 kV and greater) showed a distinct and significant decline of more than 60% during the same time period. While some decline may be associated with the decline in generating capacity additions, the magnitude of the decline suggests that factors other than generator outlets were also affected. (Table 1-2)

Table 1-1 Average Generation Capacity Additions

5-Year Period Ending	Avg. MW Increase/Year (for 5-Year Period)
1975	33,453
1980	20,685
1985	16,127
1990	13,705

Source: EEI historical statistics

Table 1-2 Average Bulk Transmission Lines Added (in the United States)

5-Year Period Ending	Avg. Net Miles Increase/Year (for 5-Year Period)
1975	6,467
1980	4,767
1985	4,051
1990	2,269

Source: EEI historical statistics

Factors Influencing the Problem

Until the 1970s, the planning and development of generating capacity additions and transmission line reinforcement were fairly straightforward. Peak loads were projected, and the most economical plan to add or reinforce capacity was implemented within a reasonable time. The increasing widespread use of air conditioning, particularly in northern regions of the United

States, introduced a necessity to cope with summer peaks and summer rated capacities as well as winter. Changes in regulatory restrictions, especially environmental ones, cost of fuel, inflation, and the impact of the rising cost of electric energy all affected utility and power pool plans for additions and reinforcements in the years following 1970.

Cost of Fuel

A striking increase in fuel costs started after 1970, precipitated by world market conditions. Fuel costs peaked in 1986 and then decreased gradually. However, levels are still far above those experienced in 1970. (Tables 1-3, 1-4)

Table 1-3 Cost of Fuel (¢/Million Btu)

Year	Coal	Oil	Gas
1970	31	40	27
1975	86	200	75
1980	140	418	220
1985	173	451	343
1990	151	332	234

Source: EEI historical statistics

Table 1-4 Composite Cost Index (1973 = 100)

Year	Cost index
1970	79
1975	132
1980	193
1985	247
1990	286

Source: EEI historical statistics

Environmental and Regulatory Actions

Rachel Carson's book *The Silent Spring* captured the public's interest in 1962 and led to an increasing awareness of impacts on the environment. However, there were some regulatory actions at the federal level prior to that time:

- 1955 The First Air Pollution Control Act
- 1956 Water Pollution Control Act

Following *The Silent Spring*, additional federal legislation was enacted during the '60s:

- 1963 Clean Air Act
- 1965 Water Quality Control Act
- 1965 Motor Vehicle Pollution Control Act
- 1967 Air Quality Act

In 1970 the United States Environmental Protection Agency (EPA) had 5,700 employees and a $1 billion budget for its first year of operation. In 1974, this increased to 9,100 employees and a budget of $8.3 billion, and by 1988, EPA had 14,500 employees. This period marked the end of federal "guidance" in the late 1960s and the promulgation of regulatory standards with accompanying enforcement of those standards.

June 30, 1975 marked an important date in regulation of air quality when all air quality (AQ) regions were required to meet federal standards. Those that didn't were forced to contend with sanctions and new regulations. Of the 247 AQ regions, 102 failed to meet the standards and were restricted in growth because no new major source of pollution could be constructed, even if the new construction met the current pollution control standards. With electric generating stations considered a major source of air pollution, this impacted the planning and construction of new generating capacity additions to match load growth. This was particularly true in those areas failing to meet federal standards.

The Nuclear Industry

Even before the oil embargo in 1974, electric generation by nuclear-fueled plants was an alternative to fossil-fueled plants contributing to air pollution. Electric utilities in the United States purchased large quantities of nuclear units in 1972 and 1973, and even exceeded the quantity of fossil units on order in the 1970s. While 231 nuclear units were ordered through 1974, this dropped to 15 after 1974; none were ordered after 1978. In addition, orders for more than 100 nuclear units were canceled between 1974 and 1982. A number of factors influenced this decline:

- The Atomic Energy Commission (later the Nuclear Regulatory Commission) was overwhelmed by permit applications and operating license requests in the early 1970s.
- Technology changes caused the ineffectiveness of standard designs in reducing licensing and permit time. In 1970 a construction permit took 19 months and an operating license took 22 months. By 1978, this had lengthened to 45 months for construction permits and 53 months for operating licenses.
- Public opinion turned against nuclear generating stations, especially after the Three Mile Island incident in 1979, despite the fact that no injuries or fatalities resulted from this incident. Some localities such as Suffolk County in New York required evacuation plans in excess of federal requirements.

These were not the sole factors affecting the costs of nuclear power plants. Inflation rates as high as 20% in the late 1970s influenced all commerce and helped to increase nuclear plant development costs from $430/kW in 1971 to $1,880/kW in 1987. This effectively curtailed construction of any new nuclear power stations in the United States. Nuclear power stations have continued to be developed in other countries, however.

Federal regulations were eased somewhat in the 1980s. The Reagan Executive Order in 1981 said in part "regulatory action shall not be undertaken unless the potential benefits to society outweigh the potential costs to society." However, new generation capacity to support the continued load growth did not keep pace. The technology of load management to reduce peak loads and total energy requirements is one of the alternatives actively pursued by the utility as encouraged by regulatory authorities and the general public.

2

Components and Relationships of Electric Load Management

Load management can be defined generically as programs and direct actions by electric utilities or other providers—resellers, generating companies, etc.—to reduce the system peak demand and energy requirements. This is accomplished by shifting consumer loads off-peak and generally reducing consumer loads through more efficient utilization of energy. It involves resellers as well as existing and future utility systems. Motivations for load management include:

- Avoiding large capital expenditures for additional generation capacity and associated transmission
- Reducing emissions contributed by existing fossil generation
- Reducing generation reserve requirements for system reliability
- Lowering the cost of energy to the consumer

The term most commonly used referring to this is *electric load management.*

There are a variety of actions and programs under electric load management, some of which are passive. With passive electrical load management, there is no direct and immediate effect on consumer loads, such as designating rates higher at system peaks and lower for off-peak times. This influences the consumer in the timing of his individual appliance or energy-consuming functions, because he has incentive to lower his total energy cost. Another passive form of electric load management would be conservation activities. These include rebates for high efficiency lighting, more efficient appliances, incentives for better insulation, technical and financial assistance under energy audit programs, and educational activities for the consumers.

Active electric load management techniques would include direct control of consumer appliances such as electric water heaters, swimming pool pumps, air conditioners, etc. The utility may also resort to voltage reduction to temporarily reduce a system peak, generally done only under adverse conditions of demand and capacity. Reduction of system losses between the energy sources and the consumer is yet another direct means of action. This may be achieved by the utility through economic load dispatch of generators, remote control of capacitor banks, sizing of conductors, etc.

Appendix C is a glossary of the most common utility terms, including load management definitions.

Components and Relationships

Before the processes connected with load management are described, it will be helpful to understand what is to be managed, together with the why, where, and how such management is to take place.

The cost of electricity (or any product) may be generally divided into three major parts:

1. **Cost of fuel (or material).** This takes into account not only the amount required to supply the consumers' wants, but also to offset the losses in the several elements of the supply system (in the electrical system, these losses are given off as heat).

2. **Cost of the electric plant (manufacturing and delivery equipment).** The size or capacity of the several elements are determined by

the "instant" demand for electric energy, whether by the individual consumer or by the community. Unlike other commodities, the storage (or stocking) of electric energy in significant amounts is not yet economically practical.

3. **Cost of the personnel.** This is the cost of the labor required to operate the system.

While all these parts are obviously interrelated, the discussion in this book will consider only the first two parts. Generally, the impact on the overall cost of supply will be greater for part 2, the factor of demand. The problem, therefore, becomes how to achieve the optimum demand that, while taking care of consumers' needs, will make the smallest practical contribution to the cost of the delivered electrical energy.

If the utility (or any enterprise, for that matter) would attempt to provide for all the possible calls for its products at the same time, it would be unable to do so. Continuous market studies are conducted, in part, to determine the preferences of the consumer. The studies also seek to determine how these preferences may be met in such a way, spread out, so that the supply elements may function as evenly as practical. The goal is to avoid high short-time peaks that would require the installation of a large generating plant to accommodate the usage peaks. This generating capacity would otherwise remain idle during the time between peaks (see Fig. 2–1).

Studies are made of the routines of households, offices, stores, factories, and other institutions to determine what probable demands will need to be supplied at any one time. Always, close attention is given to detect changes that are, or might be, occurring on the patterns developed. Seasons of the year, holidays, weather, events of widespread appeal, and many other factors (including the uncertainties contained in view of future developments) serve to make such studies somewhat complex. Yet the effort must be made if the utility (or enterprise) is not to price itself out of business.

Coincidence Factor

The peaks in the usage of electricity are referred to as maximum demands. Each unit (device, equipment, appliance, apparatus, etc.) has its own maximum demand. Where a number of such units are supplied by the same source, or circuit, the total maximum demands supplied by that circuit

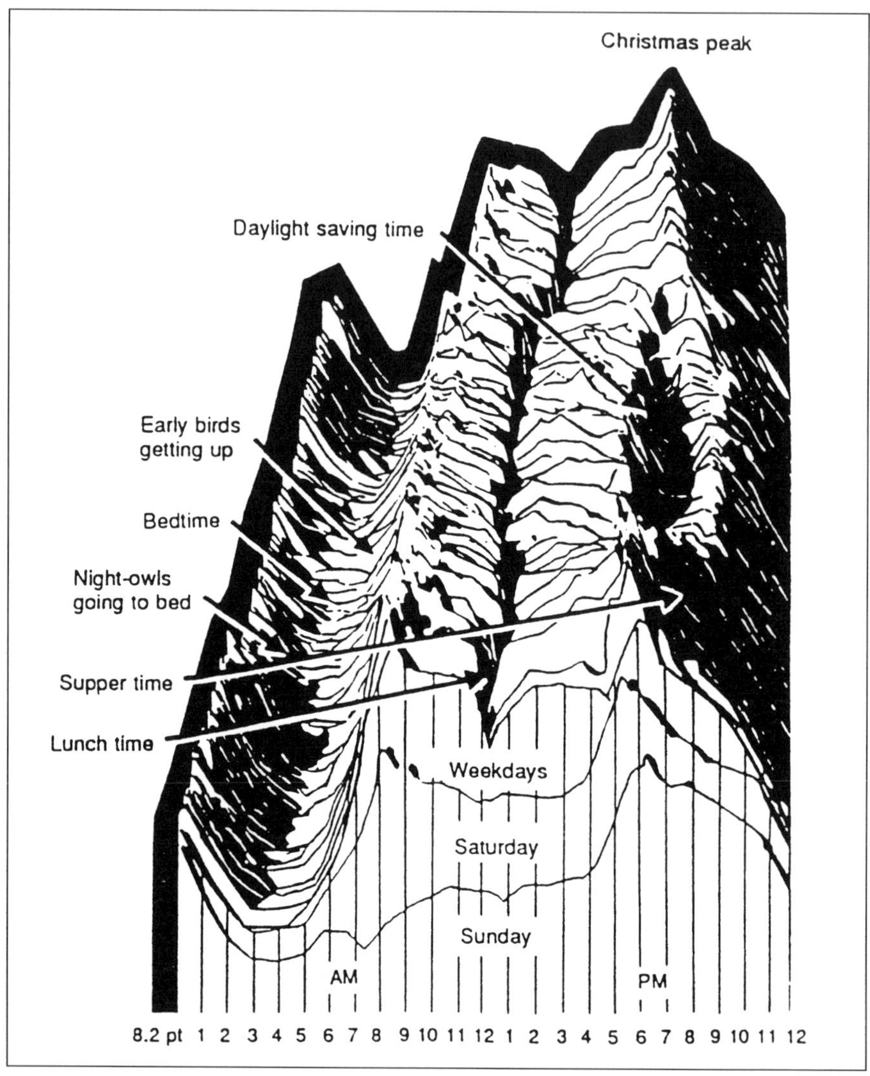

Fig. 2–1 America's routine behavior—system demand graph

is not the sum of the individual maximum demands. Instead, it consists of those demands or parts of such individual demands that happen to coincide—or the maximum demand of the circuit.

The maximum demand determines the size and type of the facilities required to meet it (see Fig. 2–2). The length of time this maximum demand may endure may vary from a very short time to a considerable time. For pur-

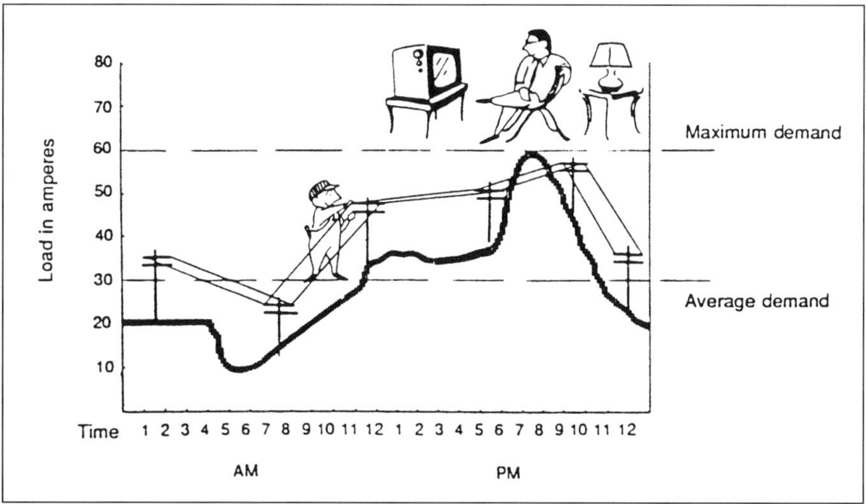

Fig. 2–2 The utility company must be equipped to meet the customer's maximum demand

poses of design (and billing), it is often taken as the highest value during a unit time period. The unit time period may be in the nature of 15, 30, or 60 minutes, occurring for a designated period of time (usually monthly, quarterly, semi-annually, or annually, depending on the purpose for its application).

Load Factor

Although the facilities installed may be determined by the maximum demand, the overall use of these facilities is an important economic factor. Thus, the facilities used for a greater length of time at or near the maximum demand are more productive (investment vs. revenue) than those used for a shorter period of time. This comparison, known as load factor, may be measured by the ratio of the average demand of all the units involved to the collective maximum demand (see Fig. 2–3):

$$\text{Load factor} = \frac{\text{Average demand}}{\text{Maximum demand}} \tag{2.1}$$

Load factors vary considerably for different types of consumers. Residential loads generally have a high evening peak and varying values at other times, depending on appliances used. Commercial and industrial loads

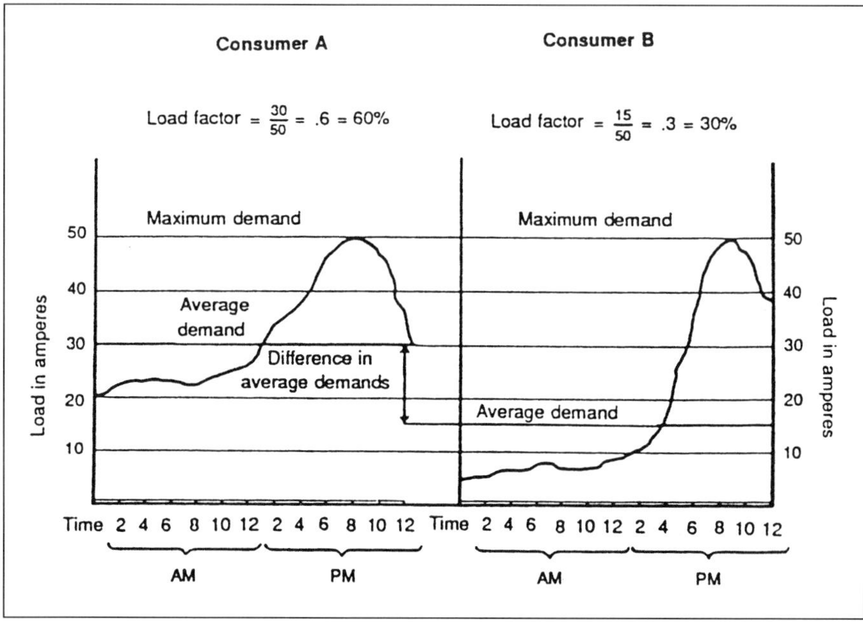

Fig. 2–3 Consumer diversity and load factor

usually have peaks of relatively long duration in the morning and again in the afternoon, while street lighting and commercial lighting loads (signs), when energized, have almost 100% individual load factors.

It is obvious that peak loads may vary for a number of reasons mentioned earlier (time of day, weather, etc.). Hence, maximum demands and load factors will vary longer than daily load cycles. Annual peak load curves (often referred to as system peaks) vary considerably less than daily peak load curves and are more often employed in forecasts and designs (see Fig. 2–4).

Use Factor

Facilities installed are not usually tailored to provide for maximum demands but consist of elements of standardized capabilities. Consequently, the ratio of average demand to the capacity of the facilities installed is used instead of the load factor. Known as the use factor, it is almost always less than the load factor.

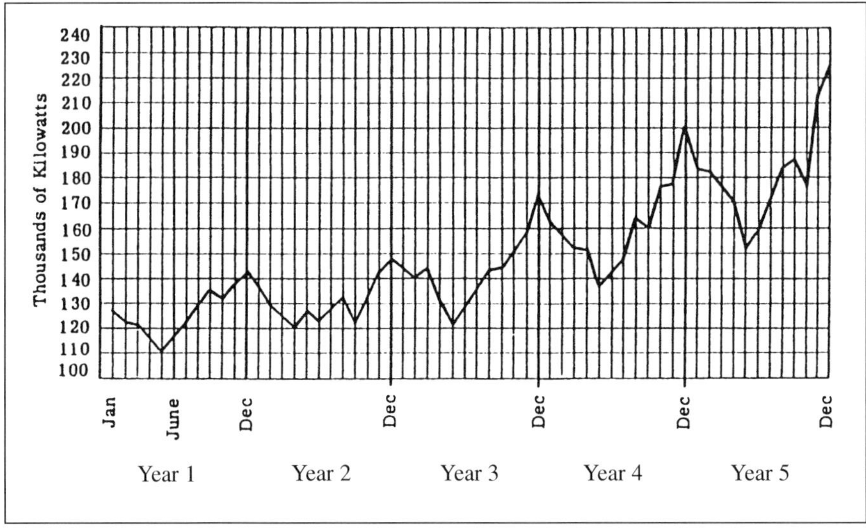

Fig. 2–4 Monthly and annual peak loads

Diversity Factor

The maximum demand on a supply source is affected by the timing of the individual demands it supplies. The source maximum demand occurs at a time when coinciding demands (not necessarily maximum demands) of the contributing units add up to produce the highest or maximum value. The relationship of the individual peak loads (or unit maximum demands) to the overall or composite peak load or system maximum demand (the measure of the noncoincidence of the individual peak loads) is known as the diversity factor (see Fig. 2–5). It is essentially the coincidence factor, mentioned earlier, that may be applied to sources of supply to multiple consumers, feeders or circuits, substations, transmission lines, and an entire utility system. It may be expressed as a ratio:

$$\text{Diversity factor} = \frac{\text{Sum of individual peak demands}}{\text{Maximum demand on supply source}} \tag{2.2}$$

Diversity and load factors are closely related. The smaller the load factor, the less chance of the maximum demands of two or more individual maximum demands occurring simultaneously. Residential loads usually have the highest diversity factor, decreasing with the increase in number of major

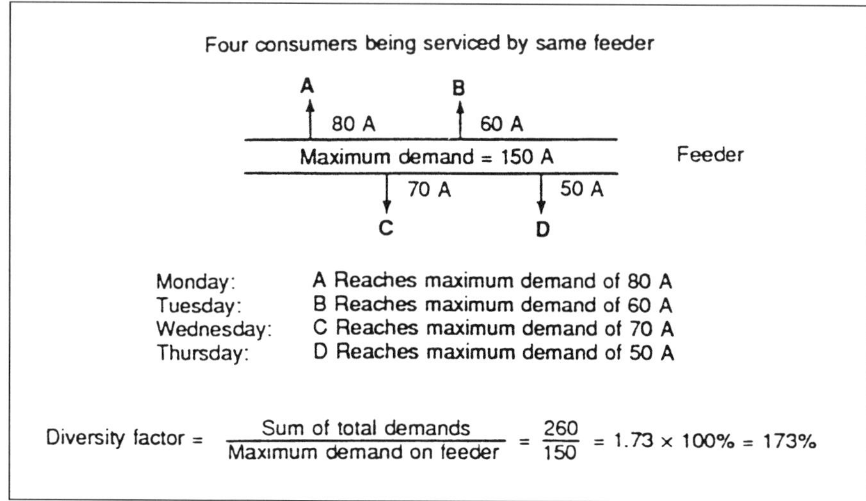

Fig. 2–5 Diversity factor

appliances used. Commercial and industrial loads have low diversity factors, generally about 1.5, and street lighting and commercial (sign) lighting about 1.0. Other loads may vary between these limits.

Summarizing, actual facilities installed provide for optimum maximum *probable* loads, rather than for maximum *possible* loads, obviously resulting in considerably reduced investment costs. The measures taken to hold down maximum demand, however, may result in less spare capacity available for contingencies in all areas, including those arising from sudden and unforeseen (e.g., military) needs.

Losses

Electricity flowing in a conductor will encounter opposition that will tend to resist or impede its flow. This must be overcome and will manifest itself in the form of a diminution in the electric motive (driving) force, pressure, or voltage, as well as in the form of heat. The loss of pressure, or voltage drop, must not be so great as to affect adversely the operation of the electrical load supplied by it. Likewise, the heat generated (which must be dissipated) should not cause deterioration, damage, or failure of the insulation, the conductor itself, or the surrounding structures. The energy consumed in the formation and dissipation of the heat constitutes a loss from the energy supplied to operate the load.

In order to alleviate the effect of the losses, it is desirable that they be measured. The relationship between the factors involved, namely pressure or voltage, current flowing, and the opposition or resistance to the current flow, may be expressed by the well-known Ohm's Law:

$$I = \frac{E_{(emf)}}{R} \tag{2.3}$$

where

I is the current flowing in a circuit, in amperes
E_{emf} is the potential difference or electromotive force
R is the resistance in ohms

The power flow (P) depends on the rate of flow of current and the pressure or voltage applied:

$$P = E \times I \tag{2.4}$$

where

P is the power in watts
E is the voltage applied
I is the current in amperes
Energy is the application of power over a period of time:

$$\text{Energy (watt-hours)} = \text{power (watts)} \times \text{time (hours)} \tag{2.5}$$

These equations enable the measurement of the quantities of the voltage drop and power and energy losses involved in direct current circuits. For alternating current circuits, however, other factors must be considered. These stem from the effect of the moving magnetic fields involved.

Electricity is produced in a wire (part of a closed or complete circuit) when it cuts (or is cut by) a magnetic field. The magnitude of the electric voltage (pressure) produced depends on:

- The length of the wire cutting the magnetic field

- The intensity or strength of the magnetic field
- The speed at which the wire cuts the magnetic field

Current flowing in a wire produces a magnetic field around that wire, the intensity of which depends on the current flowing in the wire. A direct current flowing will produce a magnetic field that will rise to its maximum strength and remain fixed. An alternating current will produce a magnetic field that will rise to its maximum strength (or value) the first quarter of a cycle, then decrease to zero in the second quarter. It will rise again to its maximum in the opposite direction in the third quarter, reducing to zero in the last quarter. The cycle repeats (see Fig. 2–6).

A wire or conductor nearby (that is part of a circuit) will have a voltage induced in it as the alternating magnetic field of the first wire cuts across it. Hence, in the second conductor, there will be two AC voltages; one is the circuit voltage, and the other is the induced voltage (see Fig. 2–7). The second AC voltage, however, will not be "in step" with the original AC voltage, but will "lag" it by a quarter cycle. This is because the AC voltage in the first conductor at its peak is not moving; hence, no voltage will be induced in the second conductor. Then as the voltage in the first conductor moves, it will be at its greatest speed as it crosses the zero line and will induce a peak value voltage in the second conductor.

The two voltages in a conductor, the original applied voltage and the induced voltage, do not exist separately but combine into one resultant voltage that will not be "in step" or "in phase" with the current flowing into the conductor. It will be seen that the power output (voltage × current) will not reach a maximum value (peak voltage × peak current), but will be only some fraction of it (see Fig. 2–8).

The percentage of this actual power to the maximum value is known as the power factor. When both the voltage and current values act together (in phase), the power factor is 100% or unity (see Fig. 2–9a). Motors with adjacent conductors carrying alternating current can cause the out-of-phase phenomenon described. (These motors are made to run by the reaction of the magnetic fields to each other.)This phenomenon may be viewed as another obstruction to the flow of electricity in the conductor in addition to the resistance (R) described previously. This obstruction is called inductive reactance. It is the sum of the effects of self-inductance and mutual inductance and is designated by the symbol X_L. The combined obstruction to the

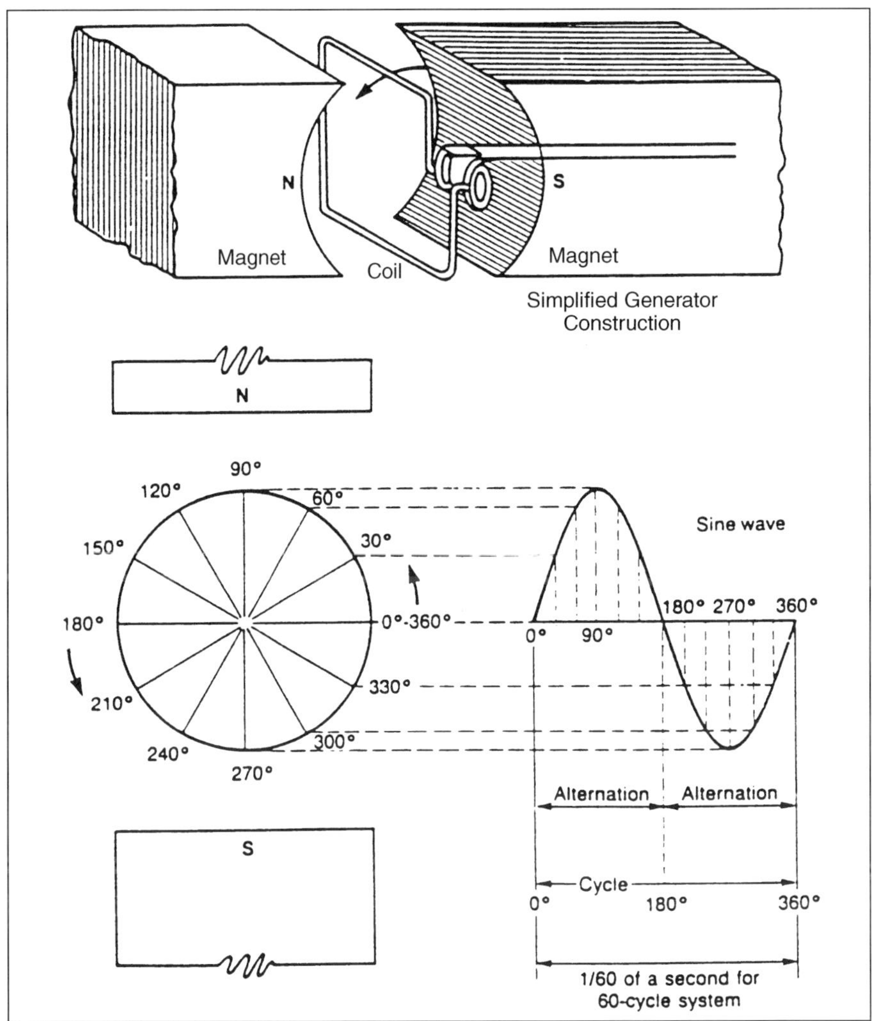

Fig. 2–6 Generation of one cycle of alternating current

flow of electricity is the effect of two "forces" acting with each other and is known as impedance, designated by the symbol Z. Its value will be given by:

$$Z = \sqrt{R^2 \times X_L^2} \qquad (2.6)$$

where

Z, R, and X_L are expressed in ohms

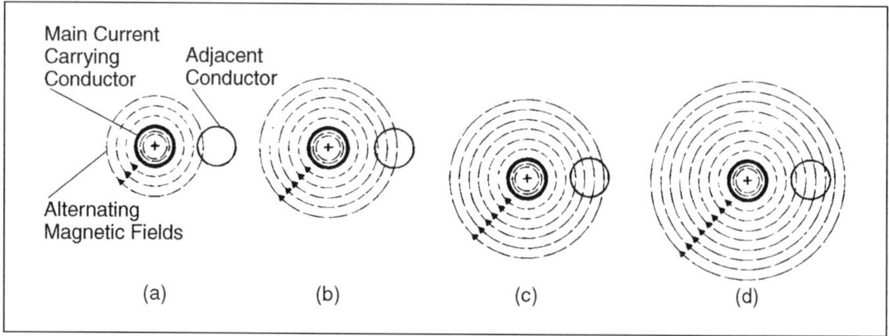

Fig. 2–7 The effect of a magnetic field about a conductor on an adjacent conductor

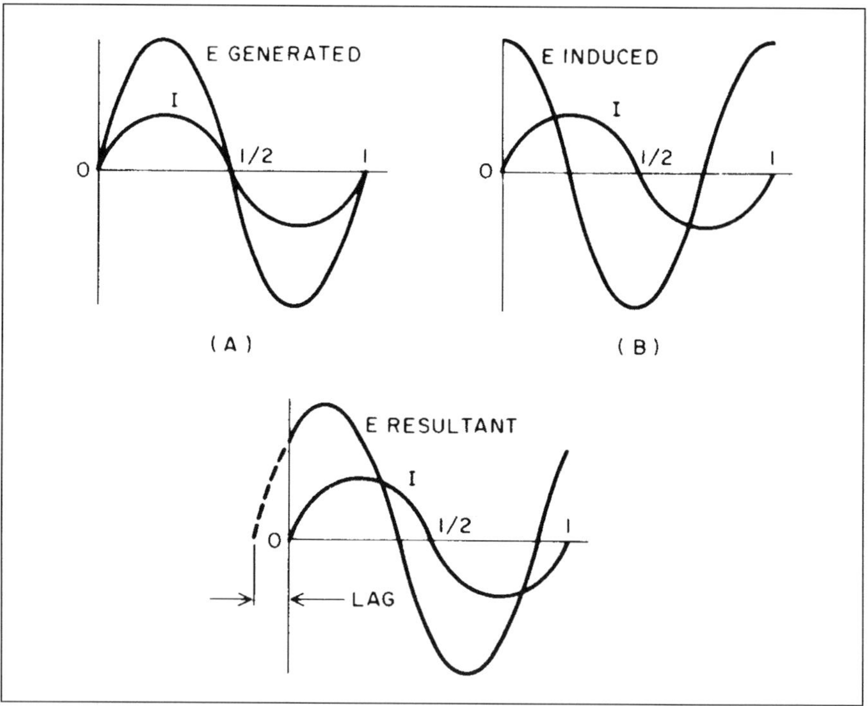

Fig. 2–8 The effect of inductance on voltage and current in a conductor

The same phenomenon exists in the conductor carrying the load current. Its magnetic field rises and falls, inducing a voltage in the conductor itself, the same as that induced by a nearby conductor coming in alternating current (see Fig. 2–10). This is known as self-inductance; that induced by the nearby conductor is known as mutual inductance.

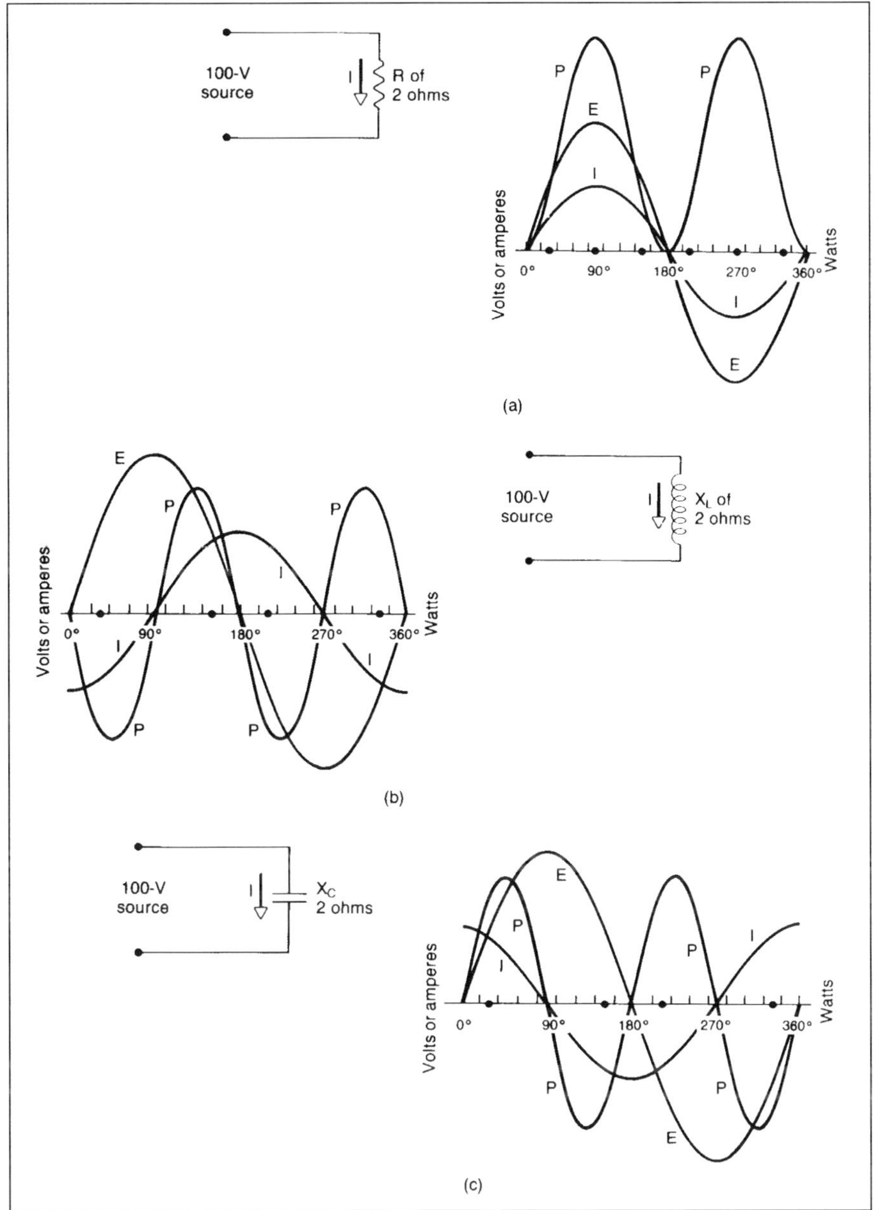

Fig. 2–9 Power values in (a) resistance, (b) inductance, and (c) capacitive circuits

Fortunately, however, there is another phenomenon in alternating current circuits called capacitance, and its effect is opposite to that of inductance. Here, the effect is electrostatic. That is, the electrons (as part of the electric current) are first attracted by the alternating current in an adjacent conductor and then repelled as the current in that conductor alternates in direction. The to-and-fro motion of the electrons then creates another voltage in the conductor, but 180° out of step (out of phase) with that created by inductance (see Fig. 2–11). Hence, the inductive and capacitive reactors tend to cancel each other. The capacitive reactance is also measured in ohms and is designated as X_C. The impedance, Z, or the total obstruction to the flow of current will be:

$$Z = \sqrt{R^2 + (X_L - X_C)^2} \qquad (2.7)$$

Capacitors are used to overcome the effect of inductance and improve the power factor of the circuit. Capacitors consist of two conducting areas separated by insulation. The value of their capacities depends on the magnitude of the parallel conducting areas, the spacing or distance between them, and the nature of the insulating material between them. The "power" effect depends on the current flow (dependent on the parallel area) and the voltage (dependent on spacing). Hence, the capacitors are rated in volt-amperes, usually referred to as reactive volt-amperes. The effect of inductance and capacitance on the non-coincidence of the voltage, current, and power curves is shown in Figure 2–9, b and c.

When the capacitive reactance and inductive reactance in a circuit are equal and cancel each other, the resultant obstruction to the flow is only the resistance of the circuit. This condition is known as resonance, and the voltage and current waves act together. The power factor of the circuit is 100%, or unity.

Let's now examine the role played by electric load management in this discussion.

Electric Load Management

Electric load management is the process of reducing the demand in all elements of the electrical supply system. Often this results in the reduction

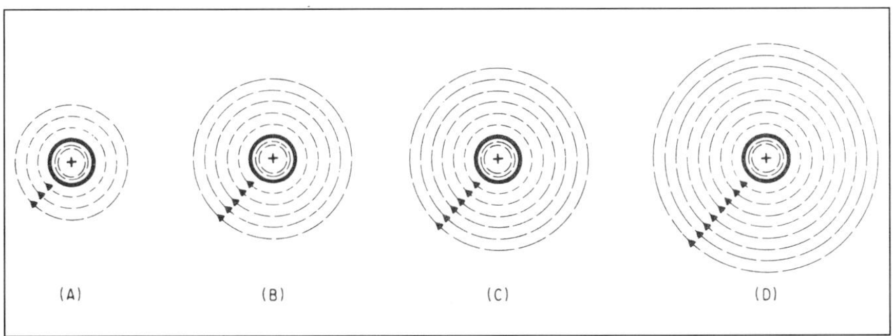

Fig. 2–10 Magnetic field expanding about a conductor (a, b, c, d) or magnetic field contracting about a conductor (d, c, b, a)

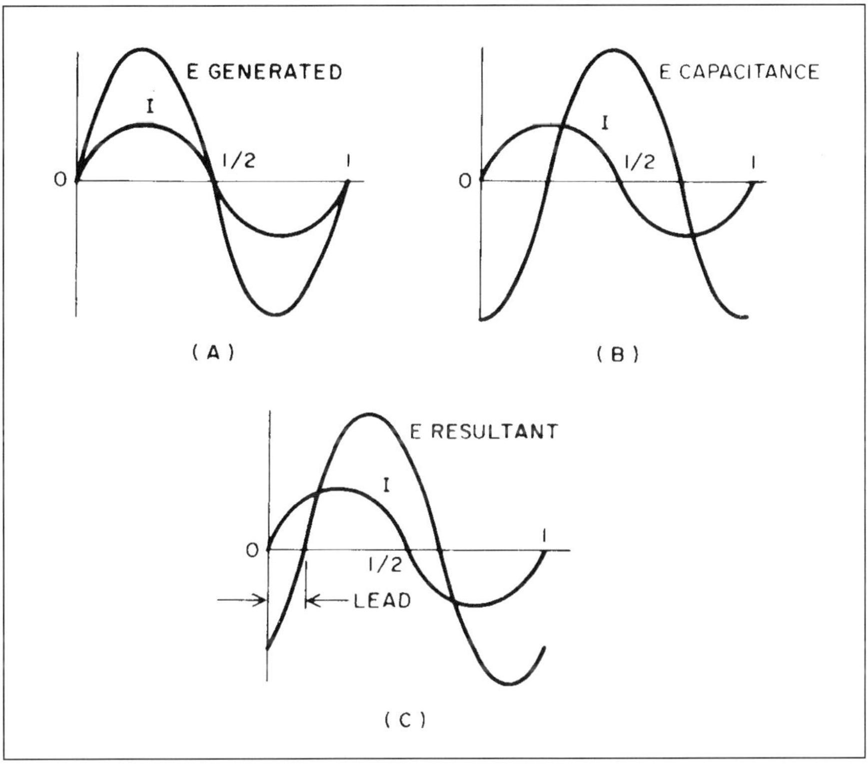

Fig. 2–11 Effect of capacitance on voltage and current in a conductor

of consumption as well. Fundamentally, there are two ways of reducing the demand:

1. Replacing the several loads with smaller and/or more efficient units (retrofitting)
2. Rearranging the time and/or duration of use of the individual units so as to avoid or limit the amount of coincidence, thus affecting the overall peak or maximum demand

The first method, retrofitting, or passive load management, is more or less self-explanatory. Replacement of units, however, should be justified economically.

The second method, or active load management, reduces the coincidence factor of the several loads contributing to the peak or overall maximum demand. As the name implies, loads may be shifted, energized and/or de-energized from time to time at the commands of controls reflecting the desired limits in demands. Controls may be programmed to operate automatically or manually, from on-site devices, or automatically from remote control centers.

While controls may be applied to all or some of the elements of the electrical supply system, it is obvious that control of the consumers' utilization of their connected loads is the basic means of reducing maximum demands. These are discussed in chapter 3.

Engineers take into consideration the impact of the reduction in demands in the several elements of an electrical supply system, as well as decreased overall consumption, in the design and operation of such systems. These include the distribution, transmission, and generating systems in integrated resource management studies. Other factors include forecasting, planning for future growth, providing for contingencies, and the maintenance of quality service as applied to proper operation of consumers' loads. Reliability, the avoidance of interruptions, and the speed of restoration are also considered.

In retrofitting, especially where units are replaced with smaller rated units, it is important to understand the basis for such ratings. Ratings of conductors, equipment, apparatus, and other devices are based on permissible operating maximum temperatures; that is, how hot they will be allowed to get. This in turn depends on the ambient conditions in which they operate, the season of the year, weather patterns, and other local considerations.

The standardization of sizes, types, and construction of the several elements comprising the entire electric system provide some slack in addition to the factors of safety included in the design. Load management, while resulting in greater efficiency by improving the use factor of these elements, generally also increases the possibility, if not the probability, of unplanned interruptions to electric service.

Even without the more detailed discussion in later chapters, it should be evident that many of the changes to reduce demand occur at the consumers' premises. Incentives to accomplish these changes on the part of the utility may include financial benefits such as out-right grants, low-interest loans, credits, and specially designed rate structures. Concentration on the larger commercial and industrial consumers not only produces larger results but may be more easily and quickly achieved. Similar measures are also productive for the smaller residential consumers.

Reiterating that changes should be economically justified, it may be well to consider them in connection with planned maintenance, revamping, improvements, or expansion of all or some of the elements involved. Indeed, it may be profitable to advance schedules of such projects to reap the benefits associated with load management.

Summary

The driving force behind load management is economics. Actual facilities installed provide for optimum maximum loads rather than maximum possible loads, obviously resulting in considerably reduced investment costs. Measures taken to hold down maximum demands should reflect optimum balances between efficiency and reliability. Such measures generally result in less spare capacity (or margins of safety) for contingencies, including sudden or unforeseen happenings (not omitting those that may be associated with military actions).

3

Consumer Utilization

The consumer, as the object of supply of electricity, can do much toward reducing the demand and consumption of that commodity. Electric load management, therefore, consists in great part of exercising control or influencing the use of electricity by the individual consumer.

Before delving into the means by which this load management can be made effective, it may be well to explore how this reduction in demand and consumption can be achieved at the consumer level. For this purpose, consumers' loads may be conveniently divided into two main categories: lighting and power loads. Other minor loads, such as computers, audio equipment, TV, and some forms of heating can be assigned to one of the major categories. The manner of management of such loads will follow.

Lighting

Lighting comprises a very large percentage of all of consumer loads. There are five different types of lamps in general use: the incandescent lamp, fluorescent lamp, mercury vapor lamp, metal halide lamp, and the sodium vapor lamp (high and low pressure). Comparative operating costs and light outputs are shown in Figure 3–l, Figure 3–2, and Table 3–1.

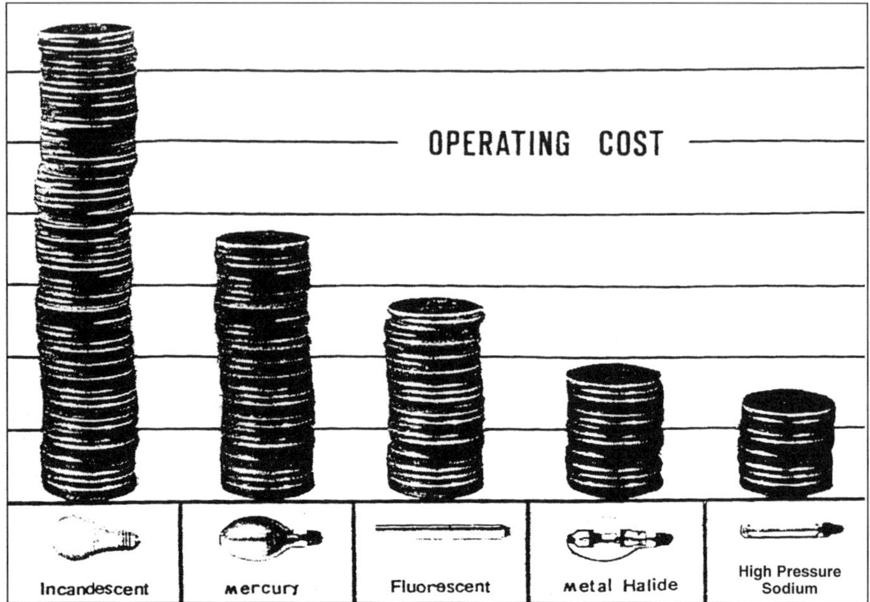

Fig. 3–1 Operating cost for various lamp types

Table 3-1 Lamp Comparisons—All Values Approximate

Lamp Type	Light Output[1] Lumens/Watt Lamp Only	Lamp & Ballast	Restrike Time in Min. Warmup	Restart	Color[2] Rendering Index	Cont. Hrs. to 90% Eat. Life[3]	400W[4] Cost Ratio
Incandescent	20	20	0	0	97		4.0
Fluorescent	100	85	0.14	0.24	60-90	25,000	20.0
Mercury Vapor	65	60	5-7	3-6	20-50	15,000	25.00
Metal Halide	125	115	2-4	10-15	65-70	15,000	45.00
Sodium Vapor High Pressure	140	125	3-4	1	20	12,000	60.00
Sodium Vapor Low Pressure	185	150	15	0	0	18,000	70.00

[1]A lumen is the measure of useful energy (light) output.

[2]Color rendering index compares with natural light taken as 100- the higher the CRI, the closer to natural light

[3]Lamp life estimate based on when lamp life output decreases to 90%.

[4]As the smallest practice high intensity discharge lamp is 400 watts, cost for fluorescent and incandescent lamps reflect equivalent.

Incandescent Lamps

Incandescent lamps are widely used, perhaps because they are the least expensive in first cost and simplicity of replacement. There are

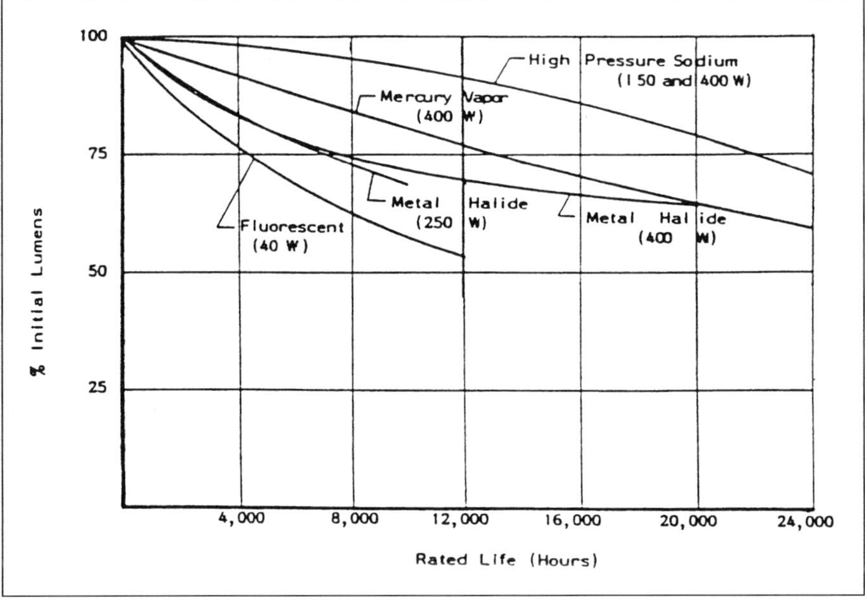

Fig. 3–2 Typical lumen maintenance curves

hundreds of different types, with variations in voltage, finish, and base. All are constructed in a similar manner and operate on the same principle (see Fig. 3–3).

In the incandescent lamp, a tungsten metal filament is suitably supported inside a glass bulb containing an inert gas or vacuum to slow down the evaporation of the metal. It is heated to a white-hot brilliance by the electric current flowing through it. The heated filament not only gives off light, but also a great deal of heat that is dissipated into the surrounding atmosphere. The glass bulb not only seals in the inert gas or vacuum, but also acts to diffuse the light from the filament.

The bulbs come in a wide variety of sizes and styles, depending on the use and voltage of the lamp. The threaded base also comes in two main sizes. A smaller or standard size is used for most lamps with ratings of less than about 1,000 watts. There is also a larger base, sometimes referred to as a Mogul, for lamps with ratings of more than 1,000 watts. Smaller bases accommodate small lamps.

The rating of the lamp is based on the current flow needed to produce the desired light output. This, in turn, determines the length and thickness

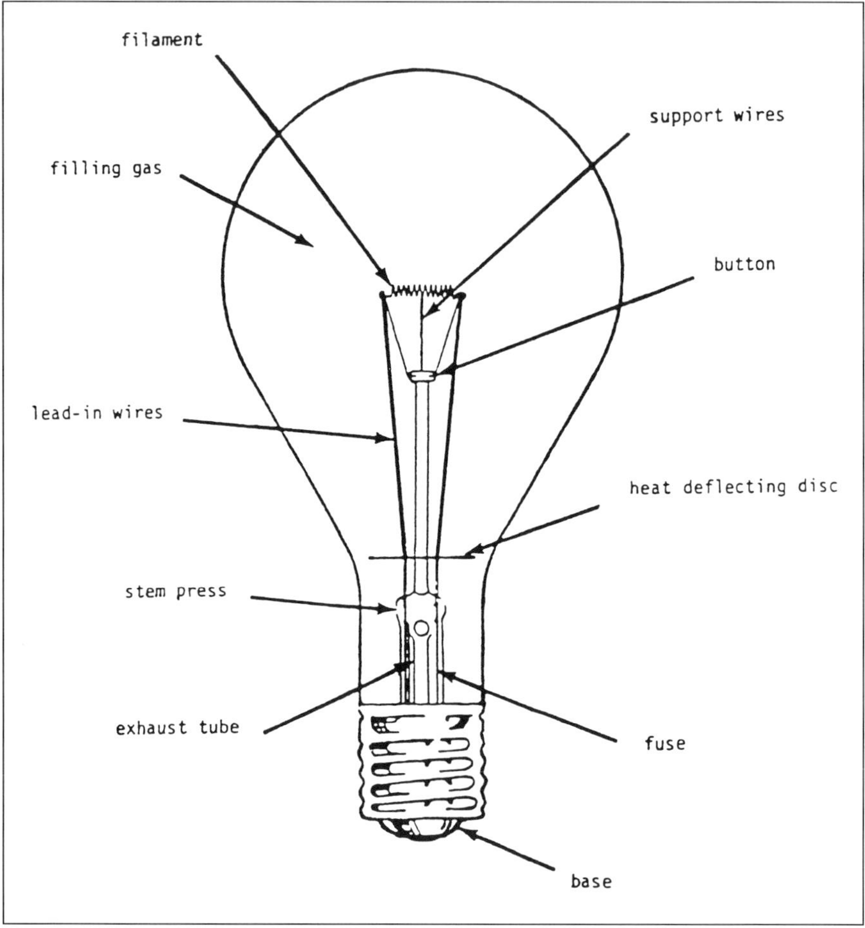

Fig. 3–3 Incandescent bulb construction

of the filament; the current flowing depending on the resistance of the filament. Although gases or vacuum in the bulb can reduce filament evaporation, the ultimate failure (or burn out) of this lamp is caused by the cycles of heating and cooling. These cycles first expand the filament and then cause it to contract. The heat also causes the metallic filament to become brittle. A weak spot develops that eventually fails, interrupting the circuit.

Incandescent lamps are generally energized very quickly by the closing of the switch controlling the circuit in which they are installed. Some switches are equipped with a built-in rheostat, or dimmer, placed in the circuit, that allows the full current to flow gradually through the filament. This decreases the strain of very rapid expansion and contraction of the metal, resulting in a greater life of the filament. Such switches for individual lamps are generally not economically justified.

The efficiency of the incandescent lamp varies widely, generally increasing with the size or rating of the lamp. For a 25-watt lamp it may be about 10 lumens per watt, rising to about 20 lumens per watt for a 1,000-watt bulb. (A lumen is the measure of light output falling on 1 ft^2 of area; a foot-candle is equal to 1 lumen/ft^2.)

The heated filament of the incandescent lamp actually emits waves of a wide range of frequencies, only a small part of which are visible to the eye. The remainder of ultraviolet and infrared frequencies represent unused energy (see Fig. 3–4).

Fluorescent Lamps

The fluorescent lamp relies on the ionization of gas molecules. These molecules act on coatings of fluorescent material on the inside of the con-

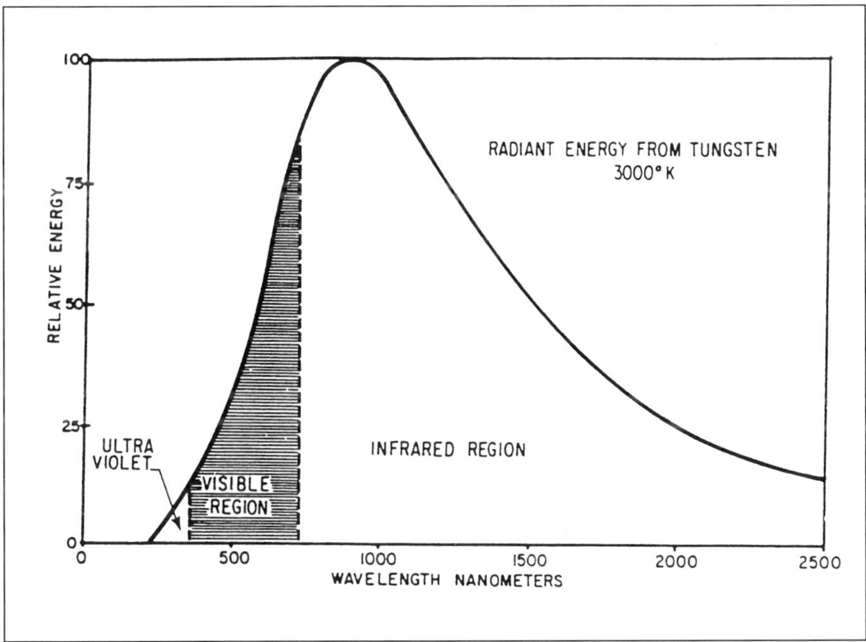

Fig. 3–4 Spectral energy distribution for 1,000-watt incandescent lamp, approximate color temperature of 1,000° K (*courtesy Sylvania*)

taining tube and produce light, but much less heat, than comparable incandescent lamp outputs.

The fluorescent lamp has two electrodes, one at each end of a coated glass tube, between which electrons flow. When energized, the two electrodes are connected in series and heated, emitting electrons. After a few seconds, a starter switch in the circuit opens, permitting the full line of voltage to be applied between the electrodes. This causes the vapor created by the heated electrodes to break down into a self-sustaining arc or discharge (see Fig. 3–5).

A ballast or reactor in the circuit limits the current to a safe value. The starting switch may be thermally or magnetically controlled and is usually contained with the ballast in a single unit (see Fig. 3–6). A drop of mercury (which vaporizes easily) is sometimes added inside the tube to aid the starting process. Fluorescent lamps may be made to shed different colors, depending on the chemical composition of the phosphor used in the coating. The light distribution for white light is shown in Figure 3–7. The fluorescent lamp comes in three basic shapes: the long tube, the circular tube, and the U-shaped tube.

Fluorescent lamps require devices to help start the ionization process described above. These are known as ballasts and are essentially small reactors placed in the lamp circuit. Their arrangement may include variations of the one described earlier. Essentially, when the lamp switch is turned on, the reactor and the high impedance of the fluorescent tube in the series allow only a small current to flow. However, it is at full voltage and is sufficient to start the arcing process. The arc reduces considerably the impedance of the

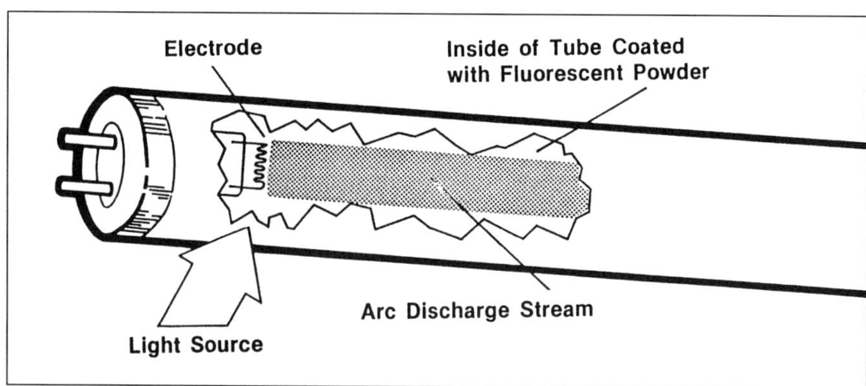

Fig. 3–5 Fluorescent tube construction (*courtesy Sylvania*)

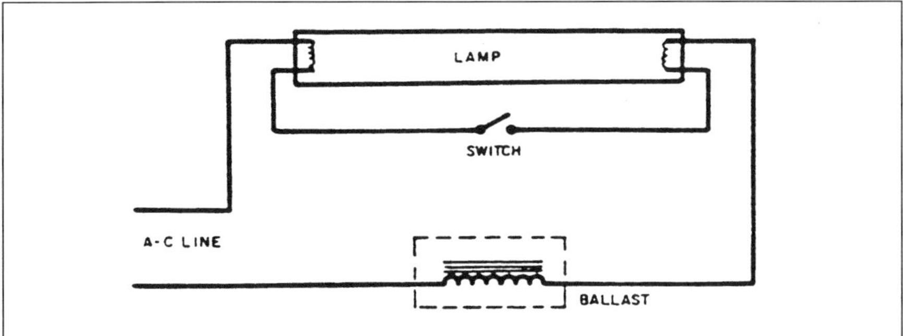

Fig. 3–6 Simple basic preheat circuit

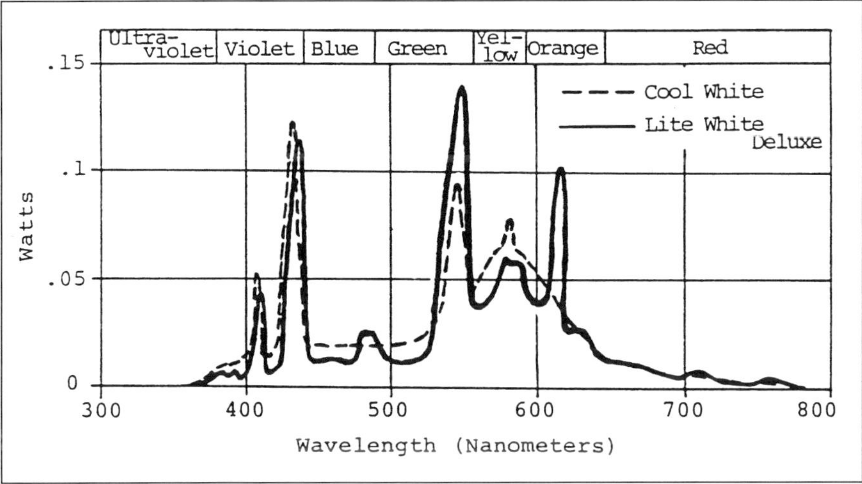

Fig. 3–7 Spectral energy distribution curves for fluorescent lamps (*courtesy Sylvania*)

tube, allowing greater current to flow, sustaining the arc. This greater current passing through the ballast reactor causes a voltage drop in the reactor. The voltage drop not only lowers the voltage of the lamp to its operating value, but also restricts the current flow to safe, steady operating values.

The installation of fluorescent lamps requires tray-like fixtures of varying shapes and sizes to accommodate the shape and size of the fluorescent lamp tubes. These lamps and their accessories are more expensive than comparative incandescent lamp installations but result in longer lamplife and greater light output. Their efficiency is about twice that of the incandescent lamps. Like them, it increases with the higher rating of the lamp, from about 25 lumens per watt to 90 lumens per watt. The efficiency and light output of most fluorescent lamps increases when the power supply frequency is increased (see Fig. 3–8).

High-intensity Discharge Lamps

Lighting for outdoor parking areas is more often served by so-called high intensity discharge lamps. These include mercury vapor, metal halide, and sodium vapor lamps, and are generally more efficient than incandescent and fluorescent lamps. In lamps of this type, light is produced by the passage of electric current through a gas or vapor under pressure. This electric current, or arc, is contained within a tube (of quartz or similar substance) that is the light source.

Mercury vapor lamps. In its simplest form, the mercury lamp is a sealed tube containing mercury, a small amount of gas such as argon, and a pair of electrodes (see Fig. 3–9). Voltage applied between the electrodes ionizes the gas and creates an arc. The heat of the arc vaporizes the mercury and causes it to emit radiation, much of which is visible as light. (Note that the fluorescent tube essentially operates in this same manner, except the fluorescent materials are deposited on the inner surface of the containing tube.)

Mercury vapor lamps usually require a warm-up period of a few minutes, in which time the current rises and the voltage drops until the arc

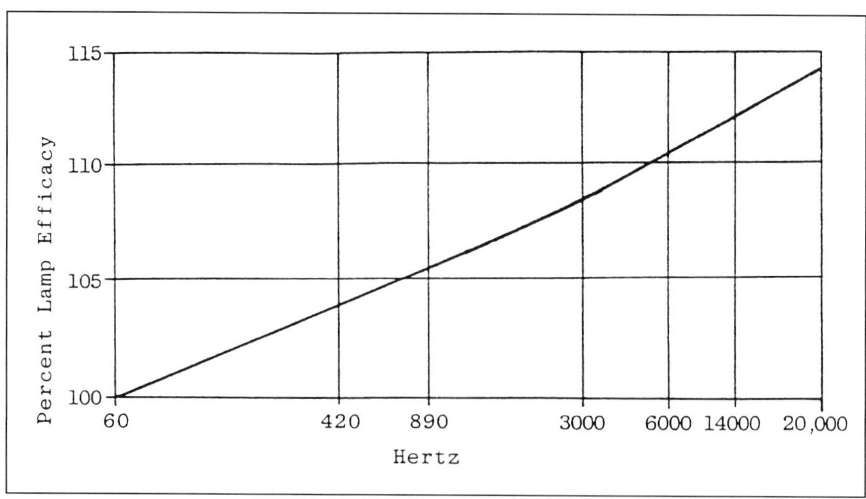

Fig. 3–8 Fluorescent lamp efficiency vs. operating frequency

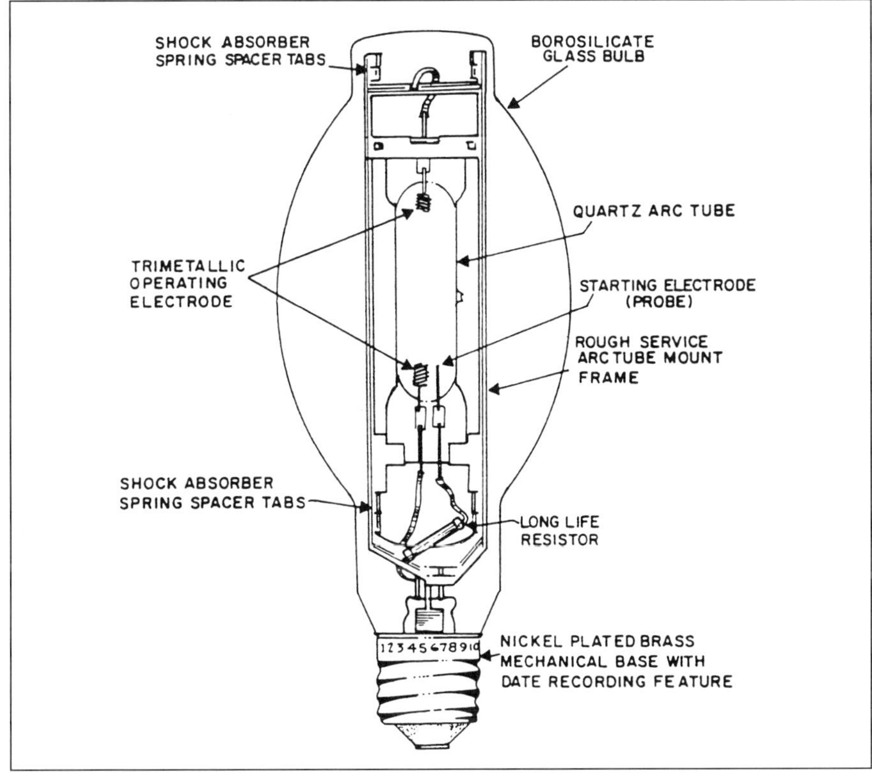

Fig. 3–9 Mercury lamp construction (*courtesy Sylvania*)

becomes steady or reaches equilibrium. The ballast here also acts to limit the current during the warm-up period and afterwards. The outer bulb, usually filled with an inert gas such as nitrogen, protects the arc tube from damage and inhibits corrosion. It also helps to regulate the arc tube temperature and eliminate ultraviolet radiation, which is damaging to the human eye. The outer bulb also deposits phosphor emissions as coatings on the inner surface that not only absorb ultraviolet radiation, but re-emits the radiation as visible light, just as in fluorescent tubes.

Long average life is a characteristic of the mercury vapor lamp, although its life output gradually declines, principally because of the deposit of emission materials on the walls of the tube. Its useful life is always longer with continuous burning than with shorter burning cycles (see Fig. 3–10).

While an efficient source of high intensity lighting, the mercury vapor lamp emits a greenish-gray color light that is not appropriate for every purpose (see Fig. 3–11).

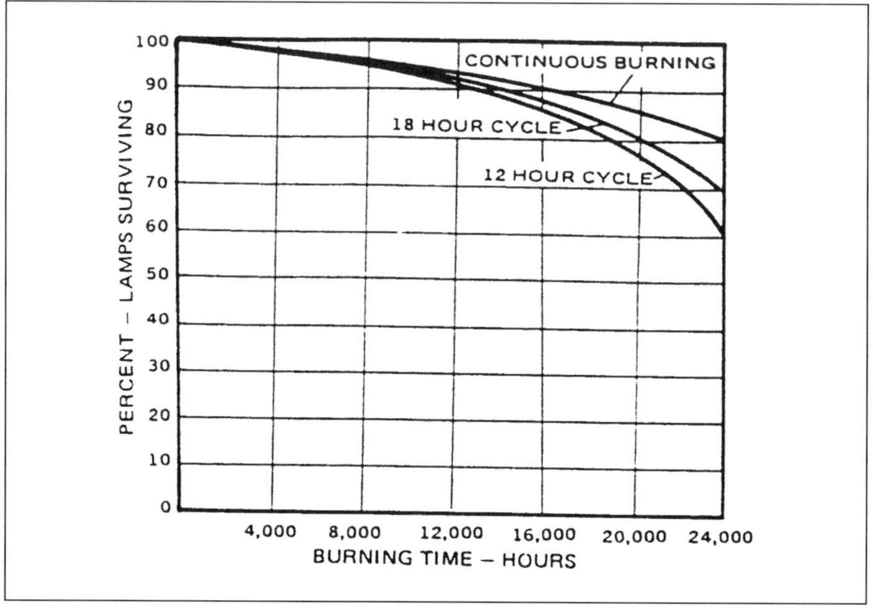

Fig. 3–10 Typical life expectancy curves for mercury lamps at various burning cycles

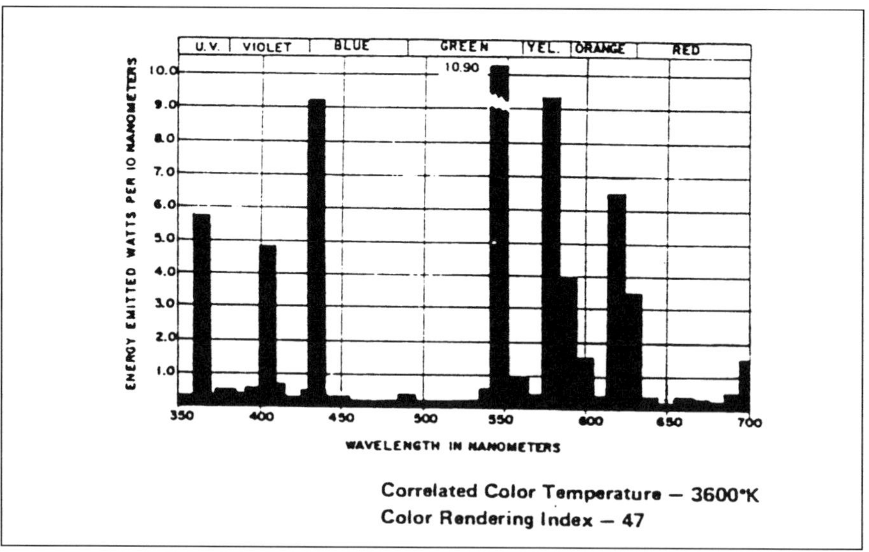

Fig. 3–11 Mercury lamp spectral radiation characteristics (*courtesy Sylvania*)

Metal halide lamps. Metal halide lamps (sometimes referred to as multivapor lamps) are high-intensity discharge lamps, essentially a modified version of the mercury vapor discharge lamp.

The metal halide lamp differs from the mercury vapor lamp in that the arc tube contains mercury vapor and argon gas, as well as additives—the metal halides of thorium iodide, sodium iodide, and scandium iodide (see Fig. 3–12). These three vapors, combining with the mercury vapor, are responsible for the white light emitted by the lamp.

The metal halide lamp employs the same starting principle as the mercury vapor lamp, with significant differences in starting requirements and characteristics. When voltage is applied to the lamp, ionization first takes place between the main electrode and an adjacent starting electrode. The metal iodides in the arc tube require a much higher starting voltage than does the mercury vapor lamp. When sufficient ionization takes place, the arc will strike between the main electrodes. Once the arc is established, the lamp begins to warm up. As the warm-up progresses, the additives begin to enter the arc stream until, when fully warmed, the lamp emits white light (see Fig. 3–13).

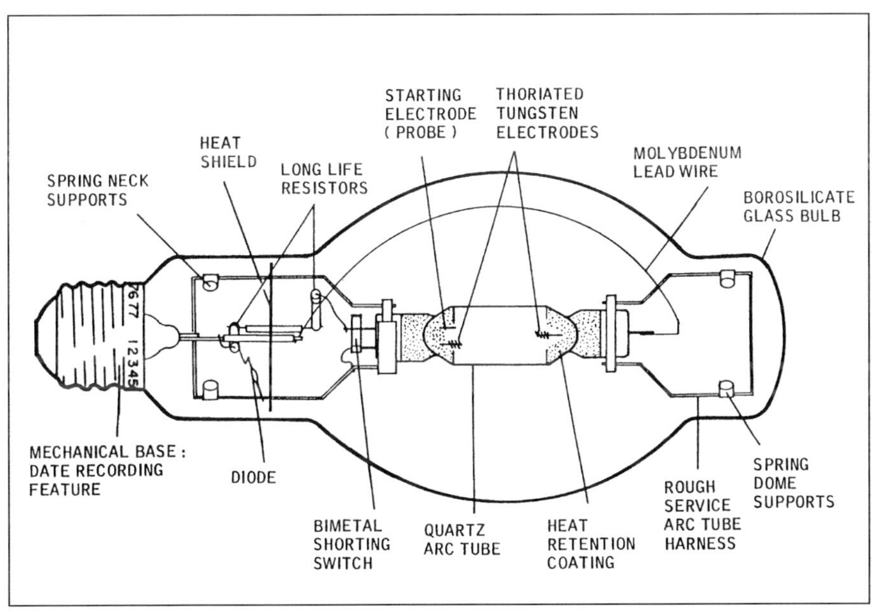

Fig. 3–12 Metal halide lamp construction (*courtesy Sylvania*)

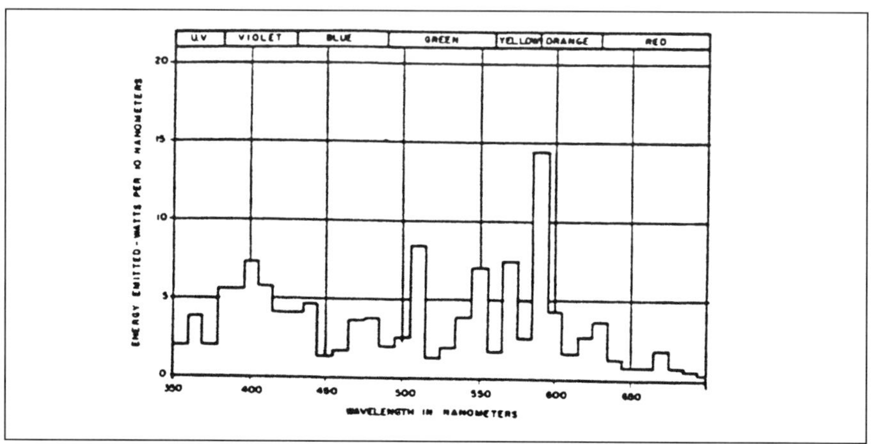

Fig. 3–13 Spectral energy emission of a 400-watt metal halide lamp (*courtesy Sylvania*)

Light output over the life of the lamp declines gradually (see Fig. 3–14). This decline is caused by the deterioration of the electrodes, the blackening of the arc tube, and the change in chemical balance of the additive material. Lamp life is affected by the number of starts (see Fig. 3–15) and by the position of the lamp (e.g., horizontal, vertical). Operation of such lamps is affected by variations in the temperature of the arc-tube wall. In positions other than vertical, the arc tends to bend upwards, affecting the

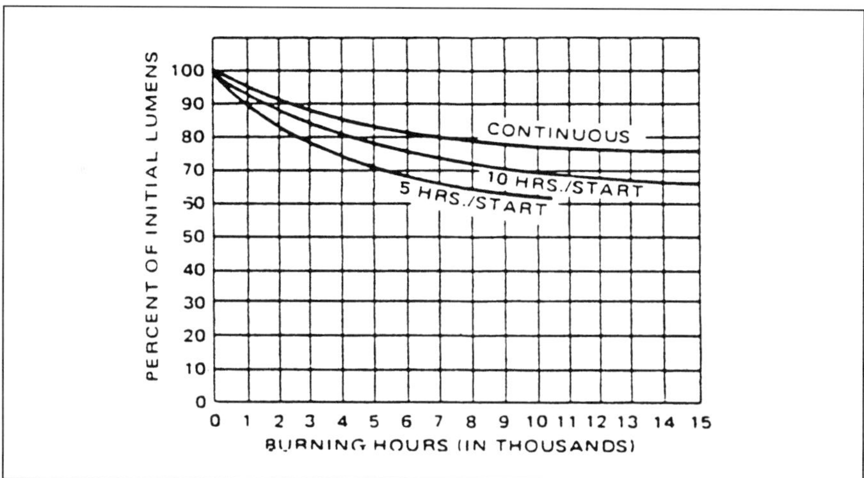

Fig. 3–14 Lumen maintenance of a 400-watt metal halide lamp (*courtesy Sylvania*)

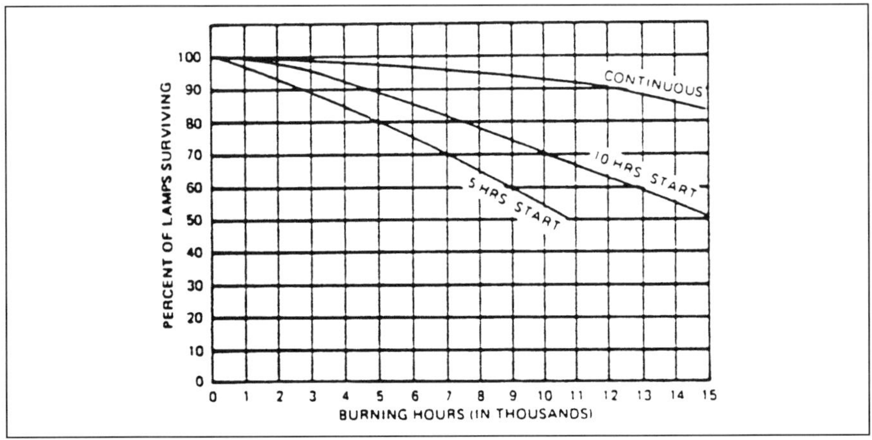

Fig. 3–15 Life expectancy curves for a 400-watt metal halide lamp (*courtesy Sylvania*)

temperature of the arc-tube wall. In a hot restrike condition following a power interruption, the lamp must cool so that the arc-tube pressure is reduced before the start-up sequence of the lamp begins.

Sodium vapor lamps. Sodium vapor lamps are the most efficient of the high intensity discharge lamps. There are two types: high-pressure lamps and low-pressure lamps. Their main advantage is that they do not emit ultraviolet radiation, as do both the mercury vapor and halide lamps.

High pressure sodium lamp. These lamps are of two-bulb construction with an outer glass container and crystalline ceramic tube (see Fig. 3–16). The ceramic material is extremely resistant to the sodium vapor and can withstand the high operating temperatures, while also being an excellent transmitter of visible light. The light produced by the high-pressure sodium lamp is a "golden white," compared to the yellow light produced by the low-pressure sodium lamp. The arc tube is filled with sodium that provides the primary emission. It also contains mercury as a buffer gas for color and voltage control, and a small amount of gas, such as xenon, to initiate the starting sequence. The arc tube also contains a set of probes; comparatively high voltages are therefore incoming for ignition. The associated starter provides a short, high-voltage pulse on each half cycle of the supply voltage that is of sufficient magnitude and duration to ionize the gas. This initiates the starting process of the lamp.

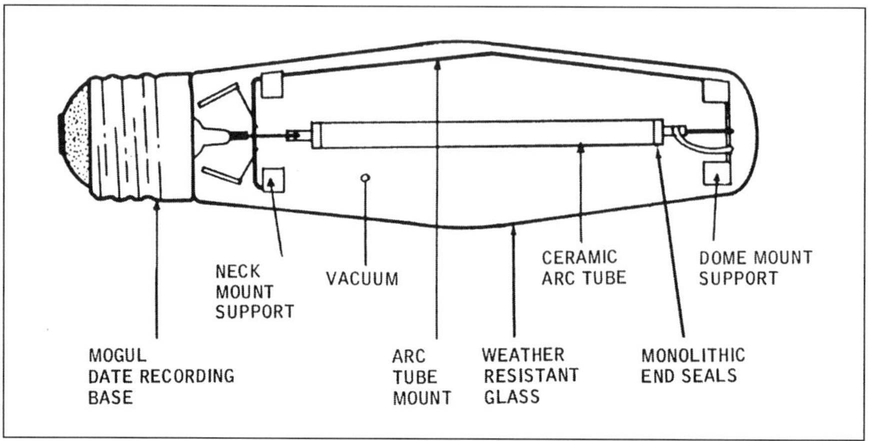

Fig. 3–16 High-pressure sodium lamp construction (*courtesy Sylvania*)

The outer container is of weather-resistant glass and contains a vacuum that helps to reduce heat from the arc tube. Problems of glare and nonuniform light distribution are solved by having an inert diffusing coating applied to the inner surface of the glass container. The coating has no effect on color, and only a slight reduction in the efficiency of the lamp. Certain high-pressure sodium lamps are designed to operate with some types of ballasts used with mercury vapor lamps and can retrofit or replace the mercury vapor lamps.

In common with other high intensity discharge lamps, high-pressure sodium lamps have a long life expectancy, which, however, is affected by the number of starts (see Fig. 3–17). The light output of these lamps gradually declines throughout their life (see Fig. 3–18). Comparison with other higher intensity discharge lamps is shown in Figure 3–19.

For reduction in demand and consumption of energy, the highest wattage lamps should be used that will provide uniform illumination with the fixtures and mounting heights that may be available. Efficiencies for higher wattage lamps are higher than for lower wattage ones.

The light produced by the high-pressure sodium lamp is a "golden white," compared to the yellow light produced by the low-pressure sodium lamp.

Low pressure sodium lamp. Low-pressure sodium lamps do not fall fully in the category of high-density discharge lamps but are more like fluorescent

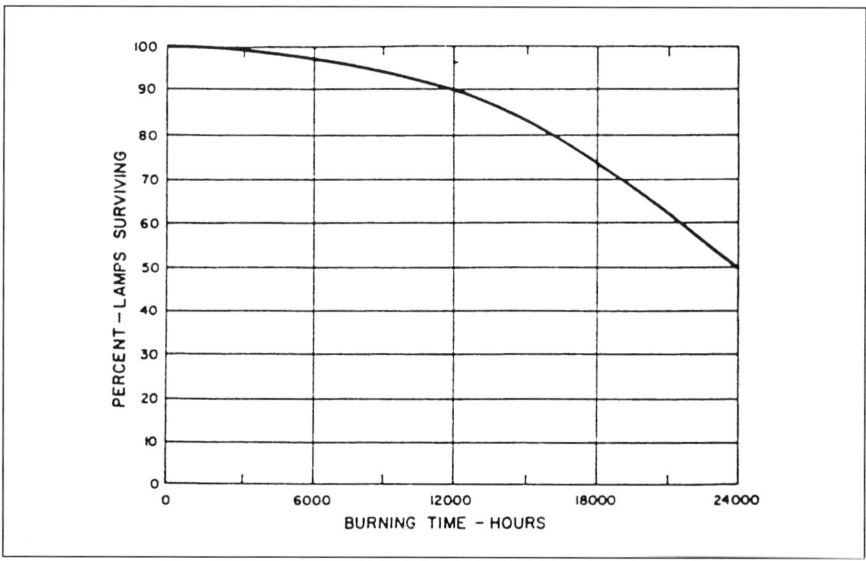

Fig. 3–17 Mortality curve for high-pressure sodium lamp (*courtesy Sylvania*)

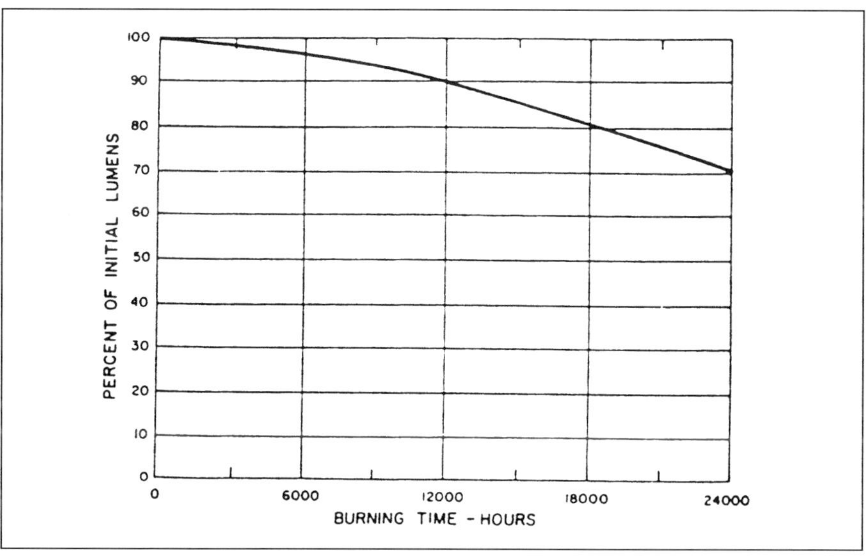

Fig. 3–18 Lumen maintenance curve for high-pressure sodium lamp (*courtesy Sylvania*)

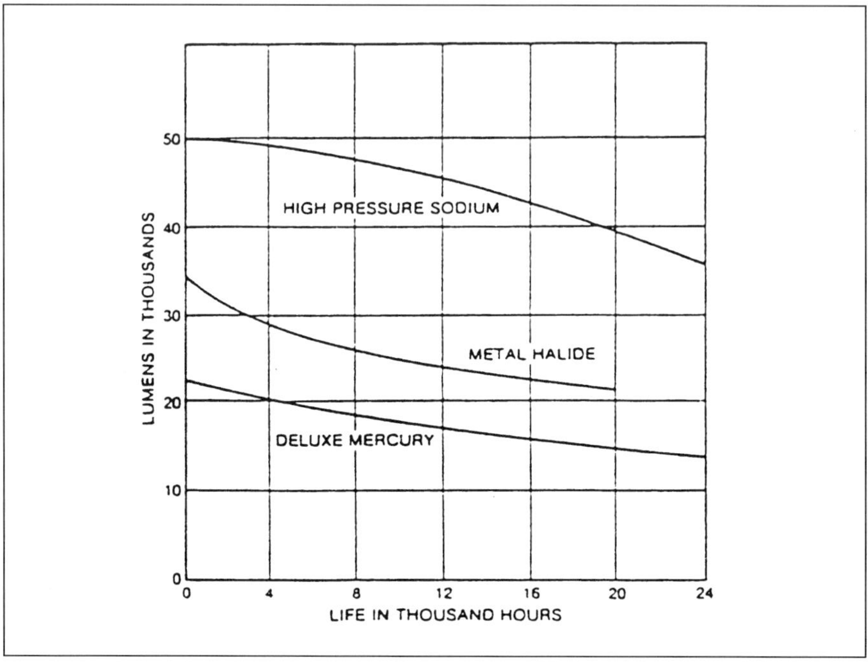

Fig. 3–19 Typical lumen maintenance curves for 400-watt high intensity discharge lamps (*courtesy Sylvania*)

lamps. Here, the inner bulb or tube is filled with sodium and a starting gas of neon with small amounts of other gases such as argon, xenon, or helium. The tube is similar to a fluorescent lamp with preheat electrodes at the ends. This arc tube, in turn, is sealed into an outer vacuum container of glass (see Fig. 3–20). Maximum efficiency is obtained when the arc tube wall temperature is maintained at about 250° C. The vacuum in the outer glass container helps to maintain this optimum temperature.

In order to obtain a high light output, a long inner tube length is desirable. This is obtained by shaping the inner arc tube into a hairpin or U-shaped tube that does not then extend the overall length of the tube. Excess amounts of metallic sodium will tend to condense at the coolest part of the tube, that is, generally at the bend. If not controlled, more of the sodium will migrate to the cool point, eventually returning the lamp essentially into a nonargon arc-type lamp. Control is established by corrugating the outer surface of the arc tube, which provides multiple cool points, ensuring even distribution of sodium throughout the tube.

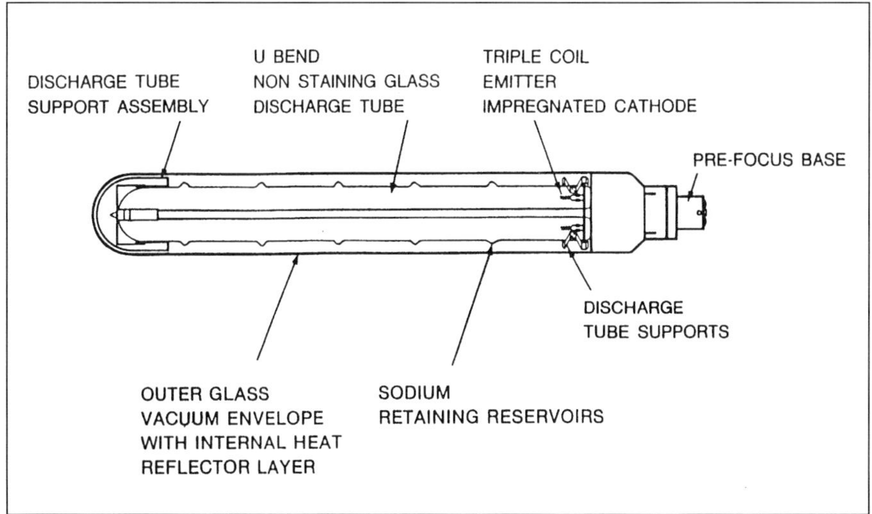

Fig. 3–20 Low-pressure sodium lamp construction (*courtesy Sylvania*)

The very high efficiency of this type lamp stems from the lack of significant radiation of wavelengths other than the narrow yellow band (see Fig. 3–21). When objects are illuminated by low-pressure sodium lamps, yellow objects appear yellow, but all other colored objects appear as shades of gray. If detection and position of objects is of sole importance, these lamps provide the most energy efficient light sources for this task. However, they will not provide identification of multicolored objects.

Full light output of these lamps is achieved in about 10 minutes after a cold strike, significantly longer than all other light sources. However, restrike after a momentary interruption is instantaneous, better than other light sources. Light output is affected by the burning position of the lamps (see Fig. 3–22). Since the losses within the bulb remain the same, the greater the wattage of the lamp, the lower the percentage becomes of these fixed losses.

Most of these lamps are used in outdoor applications. Due to their comparatively greater length, they are not interchangeable with other ballasts and fixtures. Extra caution must be taken in disposing of such lamps, as they contain metallic sodium (not amalgamates as in other high-pressure lamps). Sodium, exposed to moisture, is a fire hazard; hence, care should be used to insure the lamps are not broken.

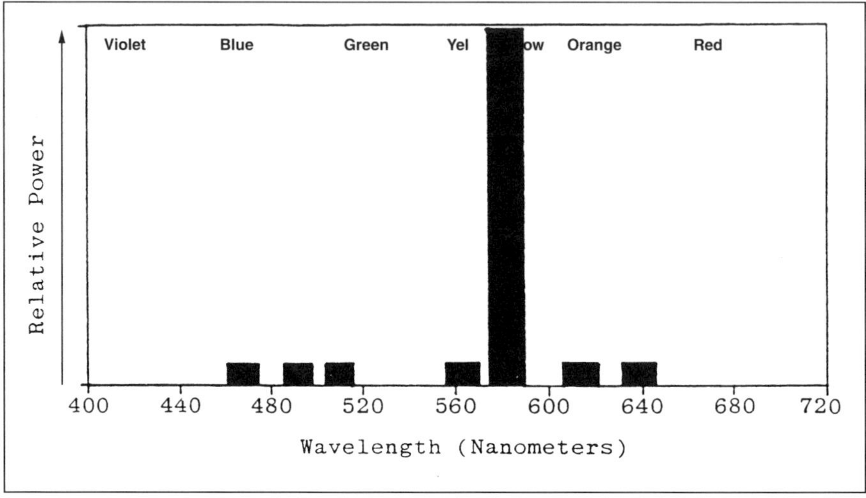

Fig. 3–21 Spectral emission from low-pressure sodium lamps (*courtesy Sylvania*)

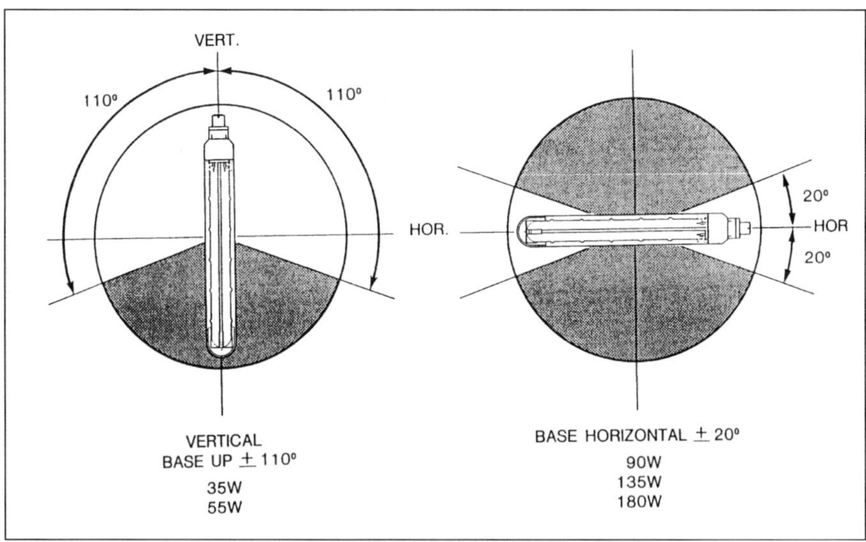

Fig. 3–22 Low-pressure sodium—burning position (*courtesy Sylvania*)

Heating, Ventilating, and Air Conditioning

Heating and cooling loads constitute a large part of the total energy requirements for most consumers of electricity. Reduction in their individual heating, ventilation, and air conditioning (HVAC) system demands and consumption would, therefore, be the obvious areas for the application of load management. (HVAC systems refer to heating, ventilating, and air conditioning equipment, and the devices and components that control their operations.) Load management of HVAC systems will tend to preserve present environments with little or no sacrifice in quality. It also recognizes the complexity of such systems and that further technical study may be required in its application to individual situations. Much of the information on this subject has been derived from publications of the American Society of Heating, Refrigeration, and Air Conditioning Engineers (referred to as ASHRAE). References to it should be made for more in-depth technical details of HVAC systems.

Basic Principles —Thermodynamic Theory

In the early nineteenth century, a French engineer, Nicolas Carnot, experimented in the field of thermodynamics and practical heat transfer, establishing the absolute limits for the efficiencies of heat engines and refrigeration systems. He concluded that the efficiency of an ideal heat engine depended on the absolute temperature of the input or source and the output or "spent" heat (sometimes called a sink); see Figure 3–23.

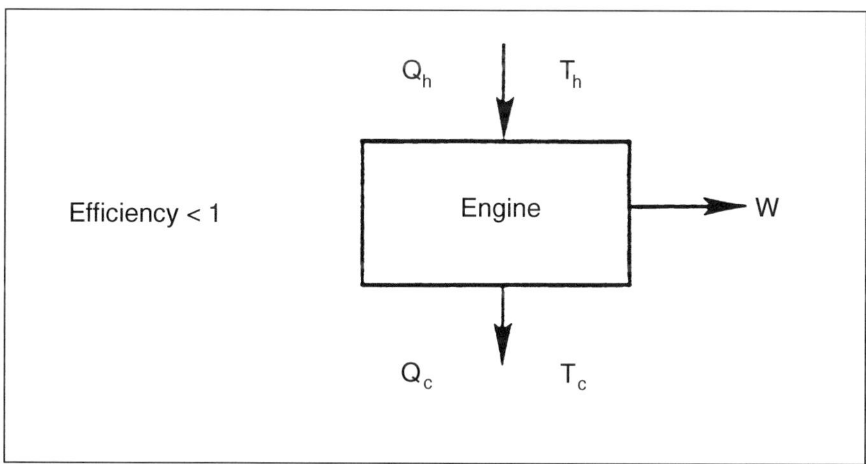

Fig. 3–23 Carnot engine

$$\text{Efficiency} = \frac{\text{Work output } (W)}{\text{Heat input } (Q_h)} = \frac{T_h - T_c}{T_h} \tag{3.1}$$

where

W is the work output in watts (horsepower or Btu/hr)
Q_h is the input heat transfer in Btu/hr
T_h is the absolute temperature of heat input
T_c is the absolute temperature of output or spent heat

It will be noted that the engine's efficiency will always be less than 100% because the heat output will always be less than the heat input. By operating the ideal engine in a reverse cycle, the ideal engine becomes an ideal heat pump. Its efficiency can be expressed by:

$$\text{Efficiency} = \frac{Q_c}{W} = \frac{T_c}{T_h - T_c} = COP_c \tag{3.2}$$

where

Q_c is the output or spent heat—usually the input to a cooling system or condenser in Btu/hr

Its efficiency is sometimes referred to as its coefficient of performance (*COP*). Because of the heat extracted from the ambient air and added to the heat input of the engine or heat pump, the efficiency rating will always be greater than 100% (see Fig. 3–24).

These equations, from a practical viewpoint, indicate that refrigeration efficiency may be improved when the incoming heat temperature T_c (evaporator) is increased, or when the outgoing heat temperature T_h (condenser) is decreased. The principles illustrated by these two equations are fundamental to the operation of a refrigeration system.

In connection with heating systems that burn fuels, the thermodynamic principles do not apply to the combustion efficiency. Such efficiencies depend on how well the heat exchanger extracts heat from the products of

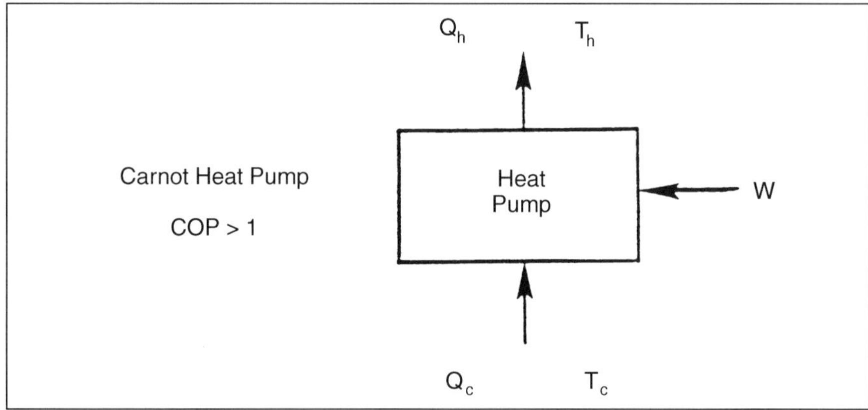

Fig. 3–24 Carnot heat pump

combustion and transfer this energy to the associated streams of air, water, or steam (see Fig. 3–25). In such indirect heating systems, 20% to 40% of the fuel energy may be carried away to the ambient environment or associated stack or chimney. When the products of combustion mix directly with the supply air, combustion efficiency may, therefore, approach 100%. Because of the obvious safety problems that arise when products of combustion are mixed with air inhaled by people, such direct-fired heat exchangers (heaters) have limited application where human comfort is concerned.

For heating units (see Fig. 3–26), total thermal system efficiency (*TSE*) can be a useful ratio. Where input consists only of fuel:

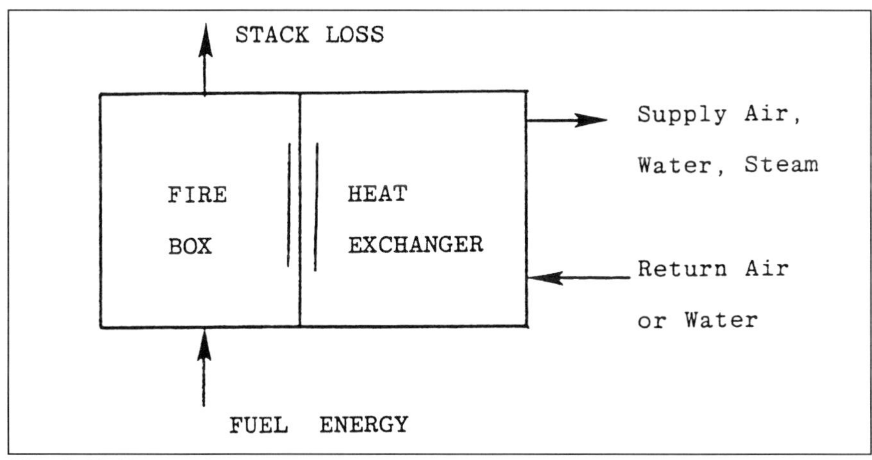

Fig. 3–25 Heat exchanger

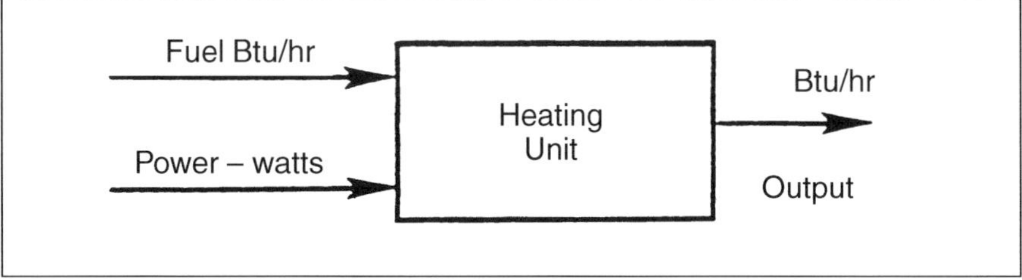

Fig. 3–26 Heating unit

$$TSE = \frac{\text{Output in Btu/hr}}{\text{Rate of total input energy in Btu/hr}} \tag{3.3}$$

Another term sometimes used is combustion efficiency (*CE*):

$$CE = \frac{\text{Output in Btu/hr}}{\text{Fuel rate in Btu/hr}} \tag{3.4}$$

For cooling systems, the standard efficiency factor is the ratio of rate of cooling output (Q_c) to the work input (rate of energy use, *W*); refer to Figure 3–27. It is known as the energy efficiency ratio (*EER*):

$$EER = \frac{Q_c}{W} = \frac{\text{Btu/hr}}{\text{Watts}} \tag{3.5}$$

This ratio is sometimes modified to take into account seasonal partial load and cycling effects, usually for residential type systems. It is known as the seasonal energy efficiency ratio (*SEER*).

Cooling system efficiency is sometimes expressed as the coefficient of performance (COP_c):

$$COP_c = \frac{\text{Cooling Btu/hr}}{\text{Input Btu/hr}} = \frac{EER}{3.413} \tag{3.6}$$

where

1 watt   is 3.413 Btu/hr

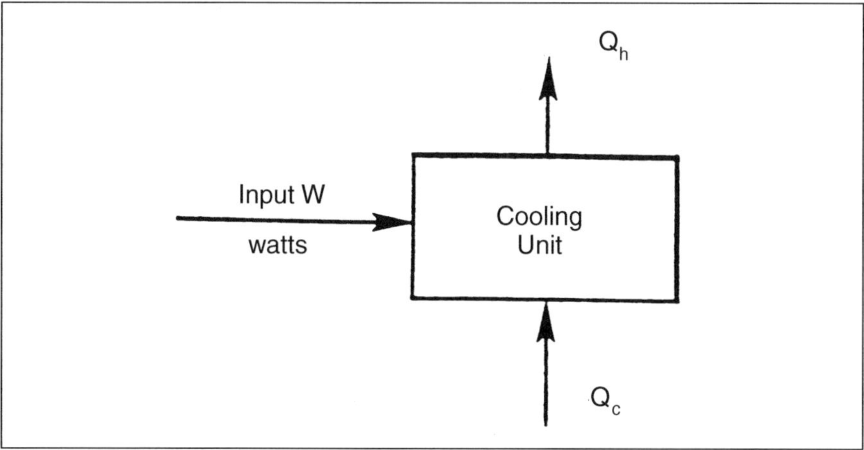

Fig. 3–27 Cooling unit

Heat Pump

The heat pump is essentially an air conditioning system with a reversing valve that allows the evaporation (heating) and condenser (cooling) to exchange functions. Its cooling efficiency is generally rated the same as an air conditioning unit, and the energy efficiency ratio (*EER*) and coefficient of performance (COP_c) are calculated as previously described. The heating efficiency is also rated in terms of the coefficient of performance (COP_h).

$$COP_h = \frac{Q_h}{Q_c} = \frac{\text{Btu/hr output}}{\text{Btu/hr input}} \tag{3.7}$$

The COP_h and COP_c are defined differently but are related:

$$COP_h = \frac{Q_{in} + Q_c}{Q_{in}} = 1 + COP_c \tag{3.8}$$

Standardized performance tests for both air conditioning and heat pump units are prescribed by the Air Conditioning and Refrigeration Institute (ARI). The ratings of *EER* and *COP* allow comparisons to be made between units of different manufacturers. Typical ARI performance data for nominal 7- and 10-ton heat pumps are summarized in Table 3-2.

Table 3-2 ARI Ratings* Cooling and Heating-MBH, Kw, EER, COP

Nominal Size	Cooling Capacity			Heating Capacity High Outdoor Temp. (47°F)			Heating Capacity Low Outdoor Temp. (17°F)		
Outdoor Unit / Indoor Unit	MBH	KW	EER[1]	MBH	KW	COP[2]	MBH	KW	COP[2]
7 Ton	93	10.9	8.5	103	9.5	3.2	59	7.0	2.5
10 Ton	118	14.7	8.0	129	12.4	3.0	76	9.4	2.4

* Rated in accordance with ARI Standards 240 and 270 MBH=1OOO's BTU ~ hour

[1]-Energy Efficiency Ratio (EER) = MBH/Total Kw input

[2]-Coefficient of Performance (COP) = MBH Output/(Total KW Input × 3.415)

Types of Heating and Cooling Systems

Components of heating and cooling systems are most commonly combined in three configurations:

1. The self-contained system, in which all components are in a single container, or single package units, including roof top units. These usually employ a heat pump *direct expansion cycle* (DX) for heating and cooling, or gas heating. Typical capacities may range from 1–50 tons.

2. The split system (see Fig. 3–28), in which the condensing coil and compressor units are installed outdoors, while the air-handling units (AHU) and heater units are located indoors. The two are connected by refrigerant piping and control lines. The same heat pump cycles (DX) for heating and cooling, or gas heating, are employed as in the self-contained system. Typical capacities may range from 5–200 tons.

3. The central hydronic system, in which a large-scale water chiller unit and a boiler are centrally located. Hot water and chilled water lines and pumps are used to distribute heating and cooling to a number of air-handling units in separate zones or buildings (see Fig. 3–29). Typical capacities may range from 200–5,000 tons.

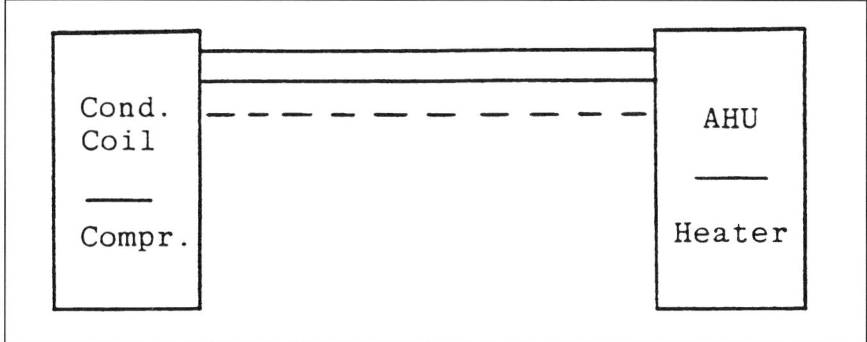

Fig. 3–28 Heating and cooling split system

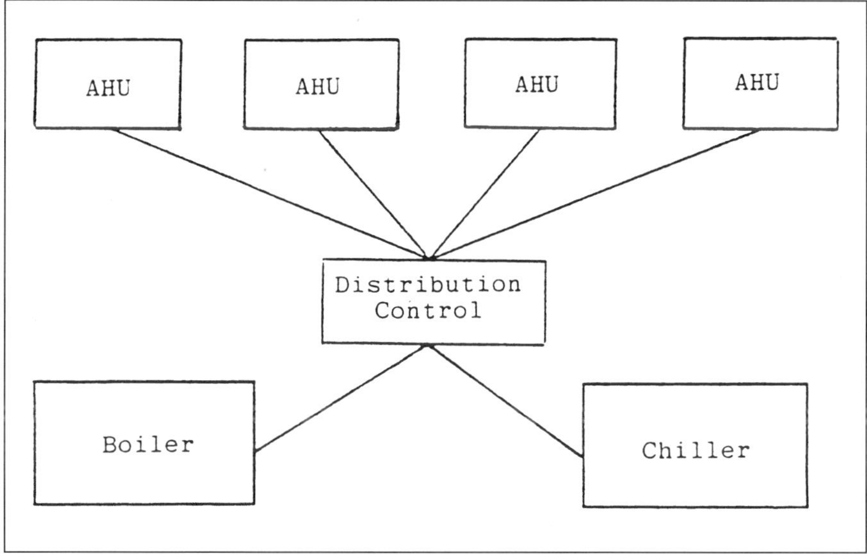

Fig. 3–29 Heating and cooling central hydronic system

Ventilation

Ventilation system efficiency (*VSE*) may be calculated from power flow rate and pressure drop data generally supplied by the manufacturers of ventilation equipment. It may be defined as the ratio of flow rate (cubic feet per minute, cfm) to input power (watts) at a specified pressure drop DP in pounds per square inch (psi). (See Fig. 3–30.)

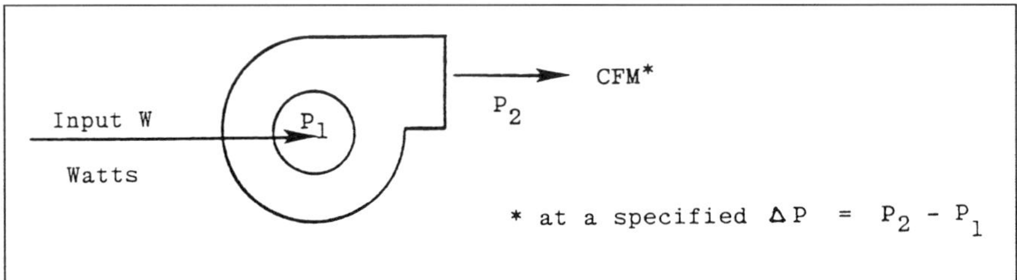

Fig. 3–30 Ventilation sytem

$$VSE = \frac{\text{Flow rate (at a specified } \Delta P)}{W} \tag{3.9}$$

where

Flow rate is in cfm
ΔP is P_1 (input air pressure) – P_2 (output air pressure)

The following guidelines may be useful in assessing the relative efficiency values of industrial HVAC equipment. (See also Table 3-3)

Table 3–3 Relative Efficiency Values of HVAC Equipment

Factor	Value	Relative Efficiency
TSE	< 65%	Low
TSE	65% – 80%	Medium
TSE	> 80%	High
*EER**	< 8 Btu/watt-hr	Low
EER	8-10 Btu/watt-hr	Medium
EER	>10 Btu/watt-hr	High
COP_h**	<2.6 Btu/Btu	Low
COP_h	2.6–3.0 Btu/Btu	Medium
COP_h	> 3.0 Btu/Btu	High

*ARI Standards 210/270
**ARI Standards 230/270 @ 47°F Ambient Temperature

Conduction Heat Transfer

The values of conduction heat transfer and thermal loads for each component of a building envelope may be calculated from the following equation:

$$Q = UA\Delta T \tag{3.10}$$

where

Q is the heat transfer rate in Btu/hr
U is the transmission coefficient of the component
A is the area for heat transfer
ΔT is the temperature difference between the inside and outside of a building

The equation indicates thermal conduction loads may be reduced by decreasing U values, reducing surface areas for heat transfer, or reducing the effective ΔT. The most common method of reducing U values in buildings is to install ceiling insulation. Wall and floor insulation may also be installed, although it is generally more difficult to install and may be inappropriate for certain buildings. Typically, U values may range from 1.1 Btu/hr/ft^2/°F for single glass windows to 0.03 Btu/hr/ ft^2/°F for roof sections insulated to a level of R-30 (see Table 3–4).

Table 3–4 Thickness Needed for R-Value Fiberglass Blanket-Type Insulation

R-Value	Thickness	R-Value	Thickness
R-38	12" (two layers of 6")	R-22	6 1/2"
R-33	10" (3 1/2" plus 6 1/2")	R-19	6"
R-30*	9 1/2" (6" plus 3 1/2")	R-13	3 5/8"
R-26*	7 1/4" (two layers 3 5/8")	R-11	3 1/2"

*Also available as single layer R-26 or R-30 batt.

Where several materials are connected in thermal series, such as multilayer ceiling sections, the U value can be calculated from the equation:

$$U = \frac{1}{R_1 + R_2 + R_3 + \ldots R_n} = \frac{1}{R_{total}} \qquad (3.11)$$

where

R is the thermal resistance of each layer in units of $hr/ft^2/°F/Btu$.

Some values of R are shown in Table 3–4, but more comprehensive updated values may be obtained from ASHRAE publications. Figure 3–31 indicates the approximate R-values that should be used in buildings at various locations across the United States.

The calculation of the optimum amount of insulation material for a building at a specific location should take into account other variable factors, such as utility rates, labor rates, weather factors, and other prevailing conditions.

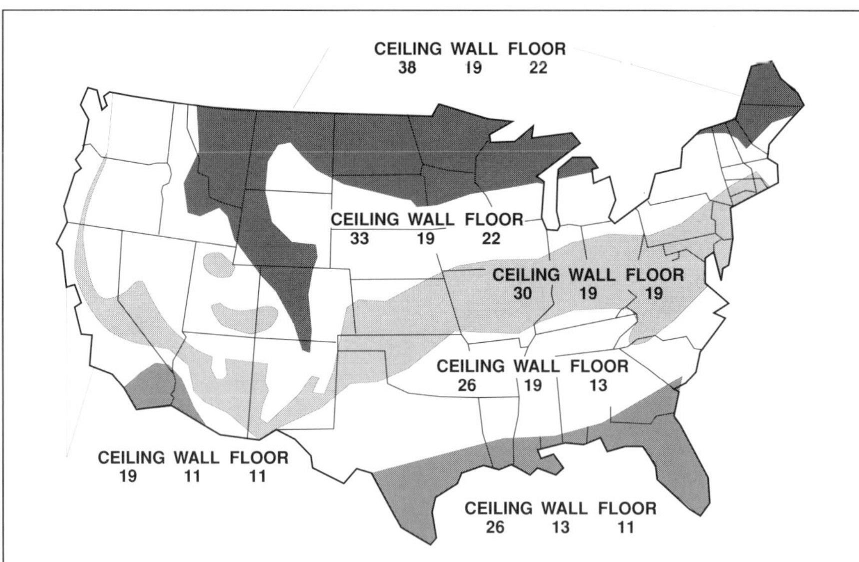

The Recommended Insulation Standards for ceilings, walls and floors shown on this map were developed from a comprehensive computer analysis of 71 U.S. cities. Owens-Corning based the recommendations on insulation costs, weather data, current and projected heating and cooling costs, and a return on investment for 20 years with savings discounted 10% each year.

Fig. 3–31 Building insulation requirements (*courtesy Owens-Corning*)

Example 3.1. Calculate the reduction of heating and cooling loads in a small manufacturing plant by adding ceiling insulation.

An energy audit of a manufacturing company revealed that the ceiling is insulated with only 2 inches of rock wool, which provides an R-value of about 5. The total ceiling area of this 20-year-old facility is 20,000 ft^2. The company operates 11 hours per day, 5 days a week, 11 months a year.

The electric DX air conditioning system has an EER = 6.0 Btu/watt-hr, and the gas heating system has a thermal efficiency of about 60%.

The optimum insulation retrofit project for this building is determined to be 10 inches of fiberglass blowing wool (an additional R-25).

The average temperature in the attic (1,848 total operating hours) is about 110° F. The attic space temperature during heating season operating hours (682 hours) is about 58° F. Natural gas costs are projected to average $9.25/mcf during the useful life of the insulation. Electricity is projected to cost $0.12/kWh during the same period.

The interior temperature of the manufacturing plant is maintained at 78° F in summer and 65° F in winter.

If the installed cost of insulation is $0.40/$ft^2$, calculate the following items:

- the reduction in cooling load
- the reduction in heating load
- the electric energy saved
- the gas energy saved
- the net annual savings
- the return on investment

Reduction in annual cooling load. The reduction in annual cooling load can be calculated from the basic heat transfer equation, Eq. 3.10:

$$\Delta Q = \Delta U\ A\ \Delta T \times \text{hr/yr} \qquad (3.12)$$

$$\Delta Q = \left(\frac{1}{5} - \frac{1}{5 + 25}\right) \times (20{,}000\ \text{ft}^2)(110°\ \text{F} - 78°\ \text{F})(1{,}848\ \text{hr})$$

$$\Delta Q = 197 \times 10^6 \text{ Btu/yr reduction in cooling load}$$

Reduction in annual heating. The reduction in annual heating load can be calculated from the same equation as above, with the appropriate ΔT:

$$\Delta Q = \left(\frac{1}{5} - \frac{1}{5 + 25}\right) \times (20{,}000 \text{ ft}^2)(65° \text{ F} - 58° \text{ F})(682 \text{ hr}) \quad (3.13)$$

$$\Delta Q = 16 \times 10^6 \text{ Btu/year reduction in heating load}$$

Electric savings. The savings in cooling and heating loads must now be translated into electricity savings:

$$\text{kWh}_{saved} = \frac{\Delta Q}{EER \times 1{,}000} \quad (3.14)$$

$$\text{kWh}_{saved} = \frac{198 \times 10^6 \text{ Btu}}{(6 \text{ Btu/W-hr})(1{,}000 \text{ W/kW})} = 33{,}000$$

Natural gas savings. The natural gas savings can be calculated by:

$$\text{NG}_{saved} = \frac{\Delta Q}{HV \times HSE}$$

where

NG_{saved} is the amount of natural gas saved, in mcf
HV is the heating value of gas (1.03×10^6 Btu/mcf)
HSE is the heating system efficiency

Substituting the available data:

$$\text{NG}_{saved} = \frac{16 \times 10^6 \text{ Btu}}{1.03 \times 10^6 \text{ Btu/bcf} \times 0.60} = 25.9 \text{ mcf}$$

Net annual savings. Calculate the ceiling insulation increase from R-5 to R-30 by installing 10 inches of fiberglass blowing wool [lowering the U value from 0.2 to 0.33 (U = 1/R)]:

First cost = ($0.40/ft^2) (20,000 ft^2) = $8,000
Annual energy saved = 33,000 kWh and 25.9 mcf
Projected rates = $0.12/kWh and $9.25/mcf
Annual savings = $3,960.00 and $240.00
Additional maintenance = $0
First cost (labor and material) = $8,000
Annual debt charge = $1,280
Net annual savings = ($3,960 + $240) – $1,280 = $2,920

Return on investment. Return on investment (ROI) is calculated by dividing the net annual savings by the initial investment cost, multiplying by 100 to express as a percentage:

$$ROI = \left(\frac{\$2{,}920}{\$8{,}000}\right) \times 100 = 37\%$$

Conduction Load Applications

One of the most effective ways to reduce conduction load is the installation of programmable thermostats that automatically raise or lower building temperature during unoccupied hours (see Fig. 3–32). By raising the

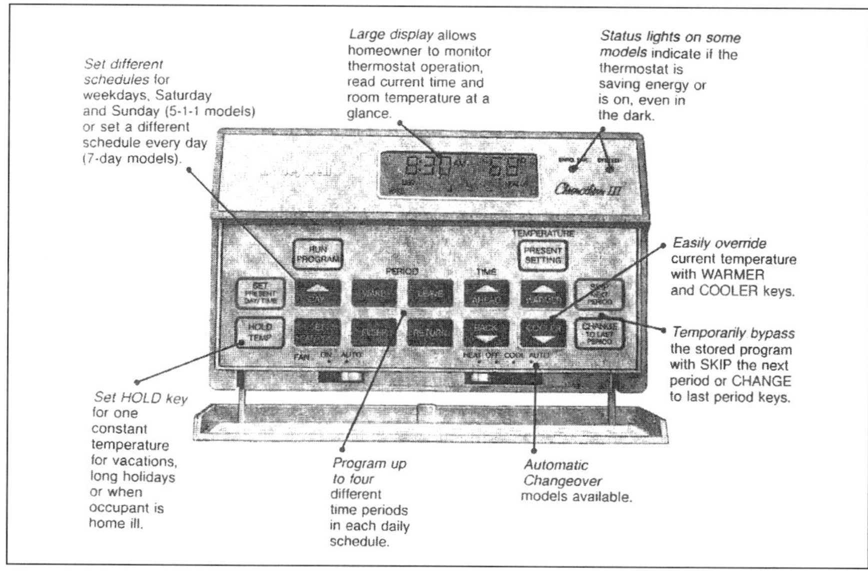

Fig. 3–32 Seven-day programmable thermostat

temperature (set up) during the cooling season, the effective ΔT is reduced. Similarly, by automatically lowering the temperature (set back) during the unoccupied hours, the effective ΔT is reduced during the heating season (see Table 3–5). Set up and set back temperature settings should be selected based on human comfort and on freeze protection limits in winter and protection of temperature-sensitive materials in the summer.

Table 3–5 Effects of Conservative Thermostat Settings on Conduction Loads

Conditions: Dallas, Texas
1,739 cooling degree-days
Average summer temperature: 81.4° F
2,146 heating degree-days
Average winter temperature: 50.8° F

Normal Thermostat Setting: Assume 72°F
Building Envelope: Assume low-mass exterior

Summer		Winter	
Thermostat Setting °F	Conduction Load Reduction	Thermostat Setting °F	Conduction Load Reduction
72	0%	72	0%
73	10.6%	71	4.7%
74	21.3%	70	9.4%
75	31.9%	69	14.2%
76	42.6%	68	18.9%
77	53.2%	67	23.6%
78	63.8%	65	33.0%
79	74.5%	60	56.6%
80	85.1%	55	80.2%

Four basic areas of application of load management include:

1. Maintenance and operating procedures
2. Thermal loads
3. Air distribution and controls (ventilation)
4. HVAC equipment

Needless to say, there are many logical approaches to load management, each with its own optimum result.

Maintenance and Operation Procedures

Maintenance and operating procedures are rarely reexamined or changed until some failure occurs. Preventive maintenance programs and scheduled periodic review of operating procedures provide opportunities for load management to achieve significant performance and economic betterment. These should take into account not only changes that might have occurred in the enterprise being served, but technological advances in all components of the system under study.

Maintenance procedures. Maintenance schedules (regular and preventive) should recognize varying intervals for such maintenance of the several components of the system or systems associated with the operation of the enterprise. For example, rotating machinery such as motors, pumps, and valves may require lubrication weekly, while screens and insulation may require only semi-annual attention.

It is important to note that maintenance often offers an opportunity to make changes in several elements of a system that might otherwise be difficult to justify.

Operating procedures. A rule basic to HVAC operating procedures is to turn off the equipment when not needed. An exception is made when reliability dictates some equipment be kept idling or spinning at no load.

Another equally important rule concerns control of the time period in which the equipment operates. The run-time for major equipment should be controlled by automatic switching devices energized on predetermined schedules. Override switches can be provided, where applicable, to provide for contingency situations.

In arranging operating schedules, coincident operation of major units, except where necessary, should be avoided. When practical, some such units should be scheduled for operating during off-peak hours, for example, during night hours. Further suggestions will be detailed in discussing particular methods and equipment.

Thermal loads

Thermal loads consist of the heat energy that must be removed from a building during the cooling season and the heat energy supplied during the heating season. This includes the heat energy lost from the building itself. The magnitude of these transfers is influenced by:

1. Heat transfer between the ambient environments inside of the building (through walls, windows, doors, etc.), including heat from the sun that is absorbed by building surfaces
2. Air flow in and out of the building due to natural and forced ventilation, wind, and infiltration
3. Internal heat generated by lights, motors, equipment, processes, and people, as well as heating equipment

These are the elements that comprise HVAC loads. Before considering electrical load management associated with these loads, it may be well to consider mitigating them by other means:

1. Install additional insulation in ceilings and walls, including retrofitting of glass windows with thermal glass or double glazing. Major manufacturers of insulation materials offer computer programs that simplify calculating economic thickness requirements.
2. Retrofit thermostats with programmable ones so that temperature and run-times are under control. Adjust temperature and run-time settings for seasonal human comfort.
3. Control heat from the sun by overhanging roof extensions where practical. Heat from the sun can also be controlled by applying absorbing screens on the outside of windows or reflective films on the interior, which may be removed in winter. Other methods include using trees for natural shading, cooling roof areas by using light-colored gravel or paint, and by thermostat-controlled roof-spray cooling systems.

Heating. Primary heating equipment includes resistance-type space heaters, heat pumps (for both air and water), heat exchangers that recover heat from sources such as air conditioners, boilers, or solar heating units. All of these require motor loads servicing fans or motors that contribute to the consumer's electric demand and consumption.

Except for some low-rise residential buildings, electricity is not generally the primary energy source for meeting requirements for both space heating and hot water heating. High-rise residential, commercial, and industrial heating requirements are generally met by the use of gas or oil. Sometimes such fuels are not available or may be less competitive with electricity. In these cases, heat pumps may be employed to extract heat from the ambient air and transfer that energy into a building's interior air for space heating. Hot water heat pumps function in the same manner, transferring that energy into the water stream.

Here, the requirements may be supplied from more than one heat pump unit, allowing for diversity and permitting the driving motors to be operated more efficiently when more fully loaded. In the case of hot water, storage tanks may be provided, allowing greater running of the motors during times not coincident with peak or minimum demands.

Heat recovery systems include devices that can be added to almost any unit where cooling or condensing is required (see Fig. 3–33). They are sometimes referred to as heat recovery generators or heat exchangers. In many cases, the heat recovered is sufficient to meet all of the hot water or warm air demands; in any event, the recovered heat may serve to preheat the water or air. The preheating reduces the burden on the associated heat pumps (or other units, such as resistance-type heaters). These devices work particularly well where there are simultaneous requirements for hot water

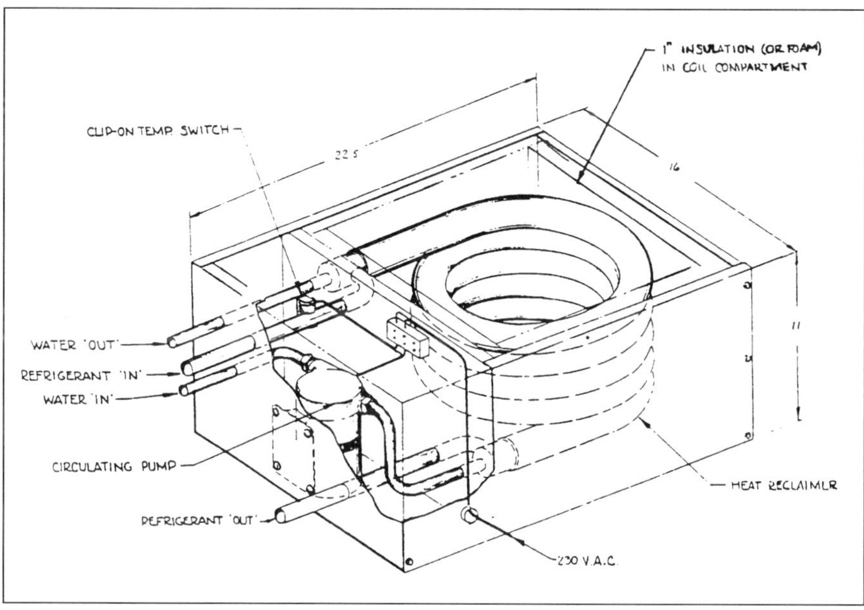

Fig. 3–33 Heat recovery generator

and cool air, such as high-rise apartments, laundries, restaurants, hospitals, and other similar facilities.

Solar heaters may also be considered for meeting hot water requirements; they are generally unsuited for space heating. These systems consist of rooftop flat plate collectors, pumps, piping, electric controls, and a storage tank (see Fig. 3–34). Due to uncertain daily weather conditions and seasonal variations in solar radiation, these systems may not be competitive economically.

Ventilation—Air Distribution

The ventilation system controlling the distribution of air in a building may provide significant reductions in the demand and consumption of electricity required for its operation. The reductions stem not only from reduced energy for moving air, but from reduced primary energy for heating and cooling. Primary energy reductions are achieved mainly by minimizing the mixing of hot and cold air streams. This is accomplished primarily through automatic devices to control damper positions and motor speed, to control air temperatures, and to connect duct work.

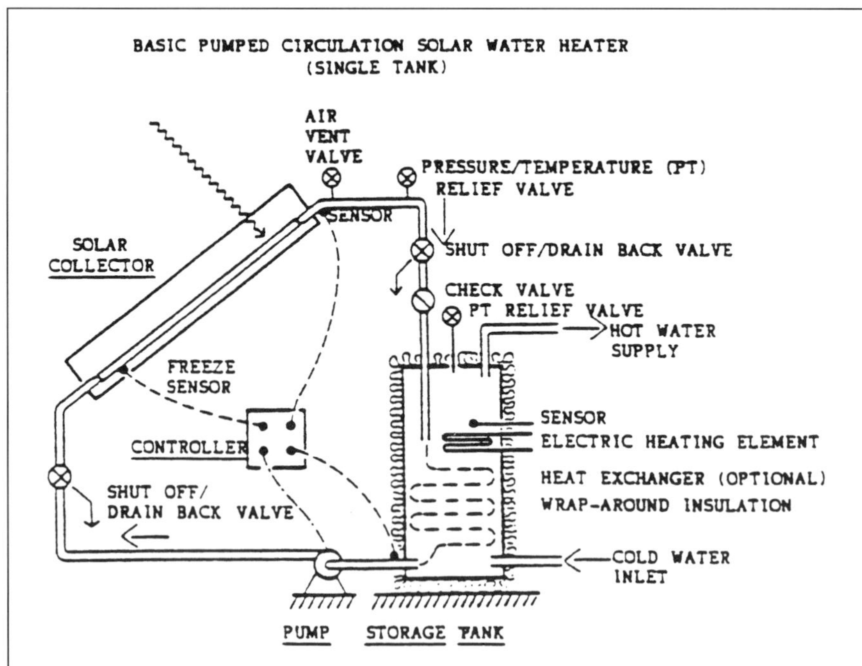

Fig. 3–34 Basic pumped circulation water heater (single tank)

Duct work to meet air control requirements depends largely upon interactions between the mixing dampers, the thermostat, and the fan. Five basic types of air control systems have characteristics applicable to different ventilation systems:

1. Single Load Control

A single zone or area is cooled or heated by a single air supply duct under the control of that zone's thermostat (see Fig. 3–35). A number of single air supply ducts service a number of "single" zones and may emanate from an air handling box (see Fig. 3–36) or one single air supply duct may have its own unit.

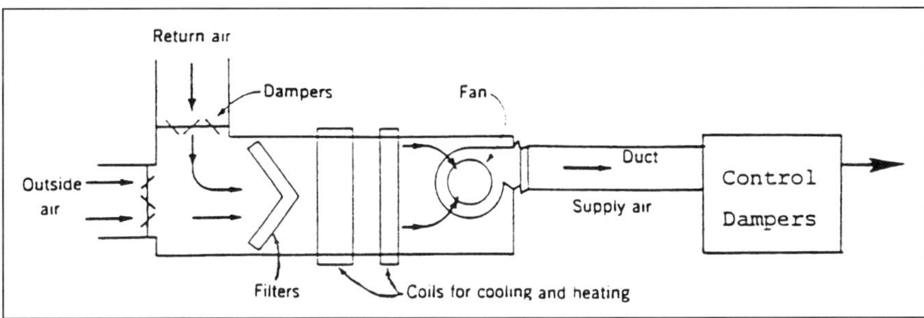

Fig. 3–35 Basic air-handling unit

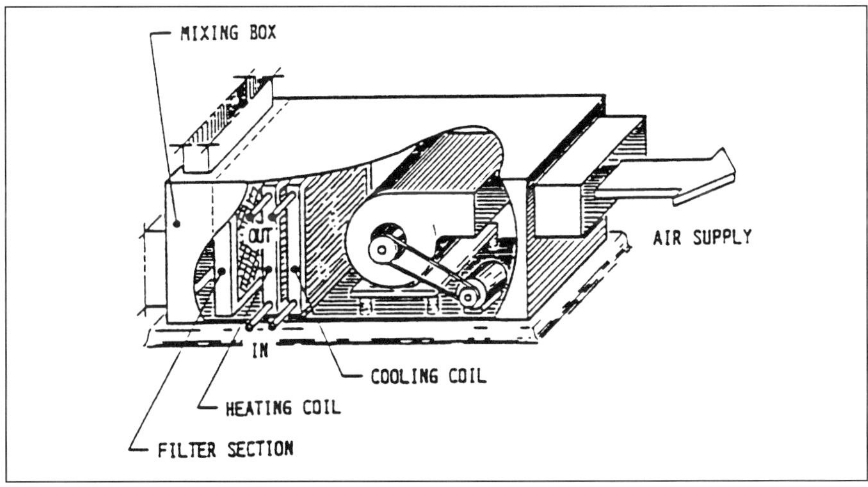

Fig. 3–36 Central station-single zone

2. Multizone Control

In this ventilation system, a single air-handling unit cools and heats several zones, each having different load requirements. A thermostat in each zone controls dampers at the unit that mix the cool and hot air to meet the different load requirements of each zone (see Fig. 3–37). Mixing takes place at the air-handling box (see Fig. 3–38), so that only one air supply duct is required for each zone. Such multi-zone arrangements are used when load levels are significantly different in each area of the building.

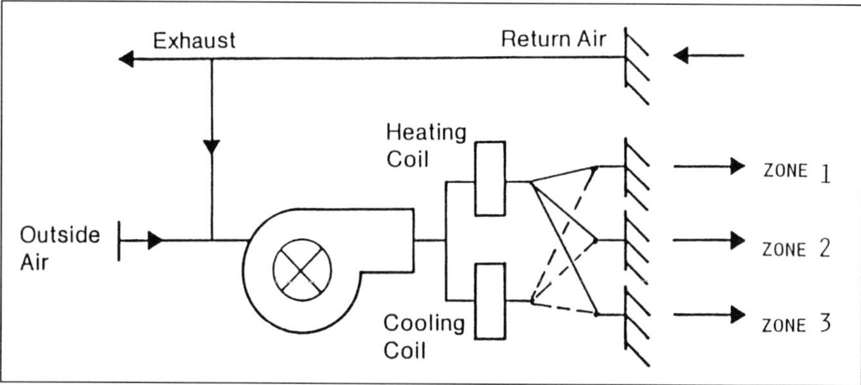

Fig. 3–37 Multizone control

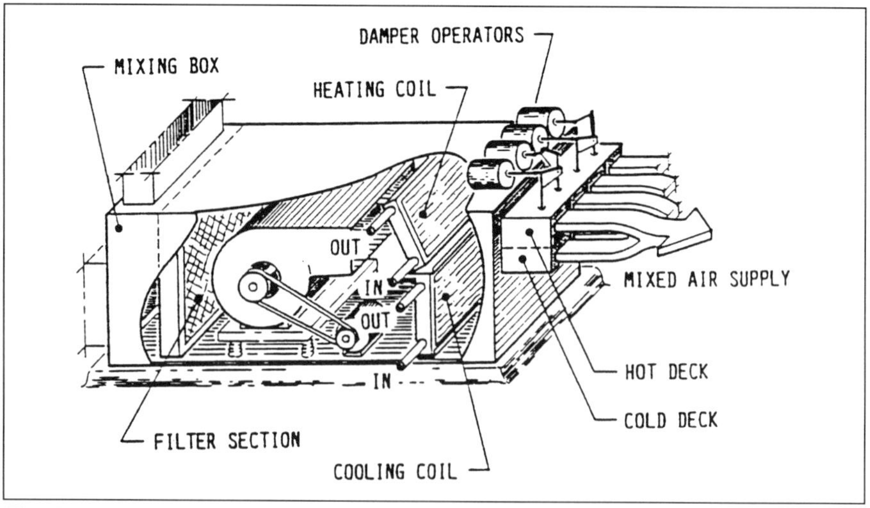

Fig. 3–38 Multizone mixing box

3. Dual-Duct Control

The central air-handling unit in this instance provides both cooled and heated air, each at a constant temperature. Two ducts serve each zone, one carrying cooled air, the other hot air (see Fig. 3–39). The ducts terminate into a mixing box where, by means of dampers, the two air streams are mixed to achieve an air temperature to meet the zone's load conditions. The position of these dampers in each mixing box (see Fig. 3–40) is controlled by a zone thermostat. Dual-duct systems applications are similar to those of multizone systems.

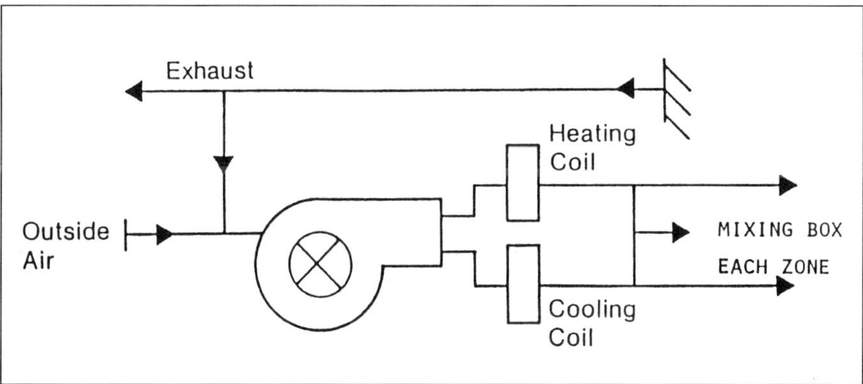

Fig. 3–39 Dual-duct control

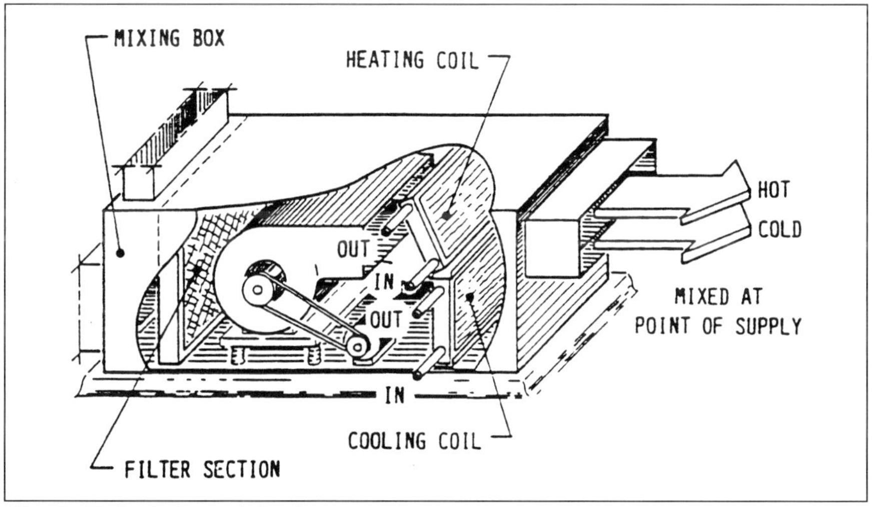

Fig. 3–40 Dual-duct mixing box

4. Terminal Reheat Control

The terminal reheat system is a modification of the single zone system in which the central air-handling unit provides very cold air to the zones served. Secondary terminal heaters reheat the air to temperatures meeting the requirements of the zone (see Fig. 3–41). While the very cold coil temperature results in excellent humidity control, the simultaneous cooling and heating of the air stream results in a relatively high consumption of electricity.

5. Variable Air Volume (VAV) Control

The variable air volume system provides cooled or heated air at essentially constant temperatures to all the zones served (see Fig. 3–42). In each zone, VAV terminal boxes adjust the quantity of air in that zone in accordance with its requirements, controlled by a zone thermostat (see Fig. 3–43). Duct work is similar to that for the terminal reheat system, except a mixing box takes the place of a terminal heater.

The system requires only single duct work to each zone, with a simple control at the terminal air unit. Since simultaneous cooling and heating of each air stream is eliminated, smaller capacity heaters and coolers may be used. As the volume of

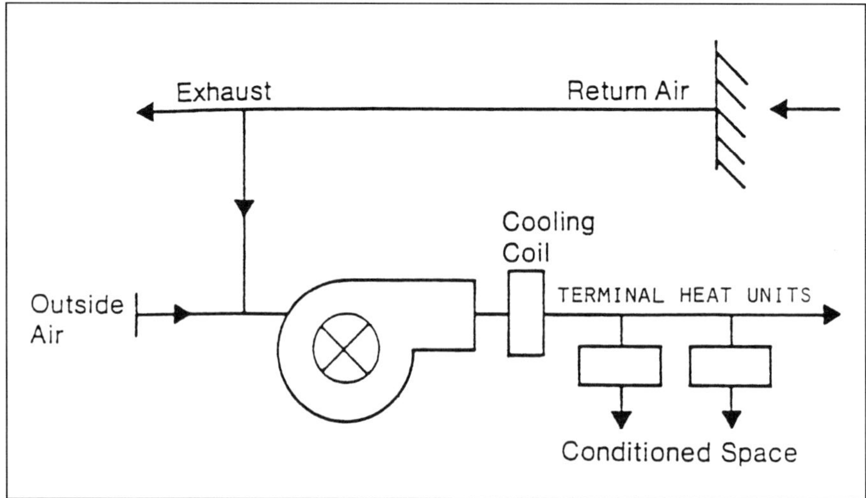

Fig. 3–41 Terminal reheat control

air can be reduced with a reduction in load, the cooling and fan power requirements can also be reduced. Hence, operating costs as well as initial costs are generally lower than the other air-control systems described.

Fans and ducts. When single speed centrifugal fans are used in air-handling units, means may be provided for reducing the flow of air to accommodate

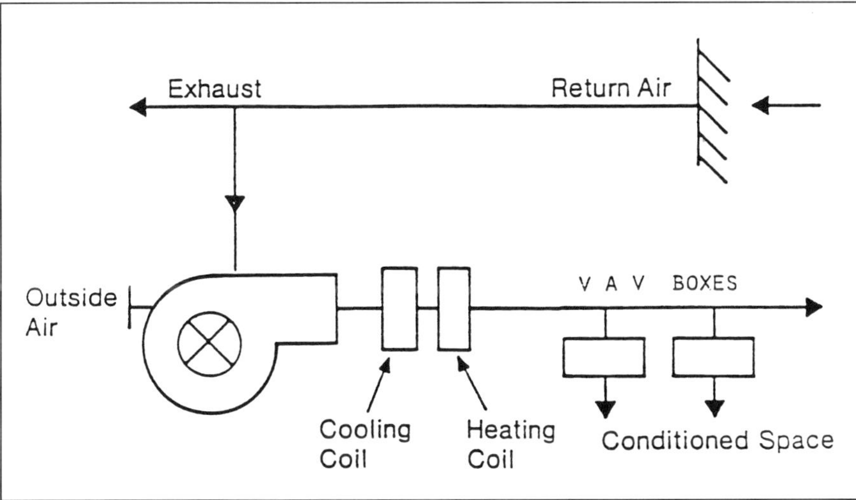

Fig. 3–42 Variable air-volume control

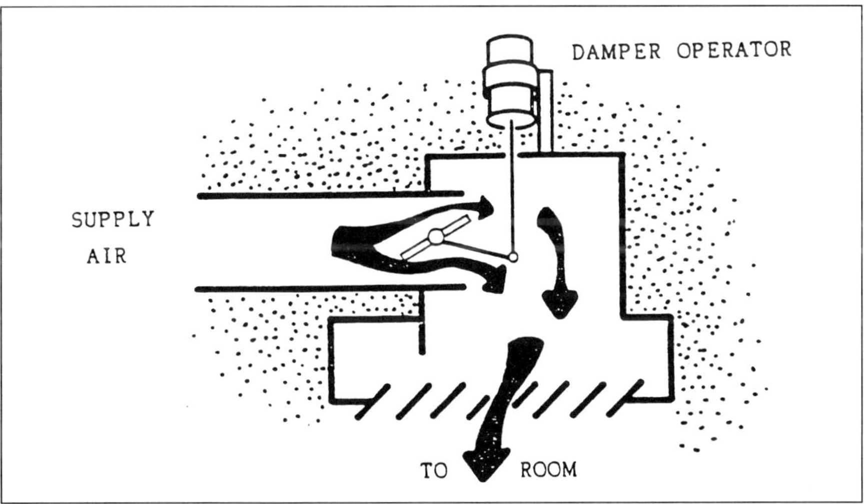

Fig. 3–43 Non-bypass VAV system

the variable requirements. Changing ratio mechanical drives that couple the fans to the motors may (approximately) match fan speeds to the variable demands in airflow. Another method is to replace the centrifugal-type fan with one having a variable blade pitch feature. Energy to drive fans varies approximately as the cube of fan or air speed. Hence, a small reduction in fan speed or airflow can result in significant reductions in electricity demands and consumption.

Electrical requirements may also be reduced by reducing the resistance to airflow in the ducts. One effective way is to eliminate sharp bends in duct systems or install turning vanes in sharp (e.g., 90°) turns (see Fig. 3–44). These methods reduce pressure drops in duct bends, smoothing the flow of air and preventing excess turbulence that causes high back pressure. Larger duct sizes also reduce resistance to airflow.

Economizing cycles. Modifications of the airflow system are made to take advantage of available air temperatures. There are usually certain times in the year when outside air conditioners are suitable for ventilation directly from

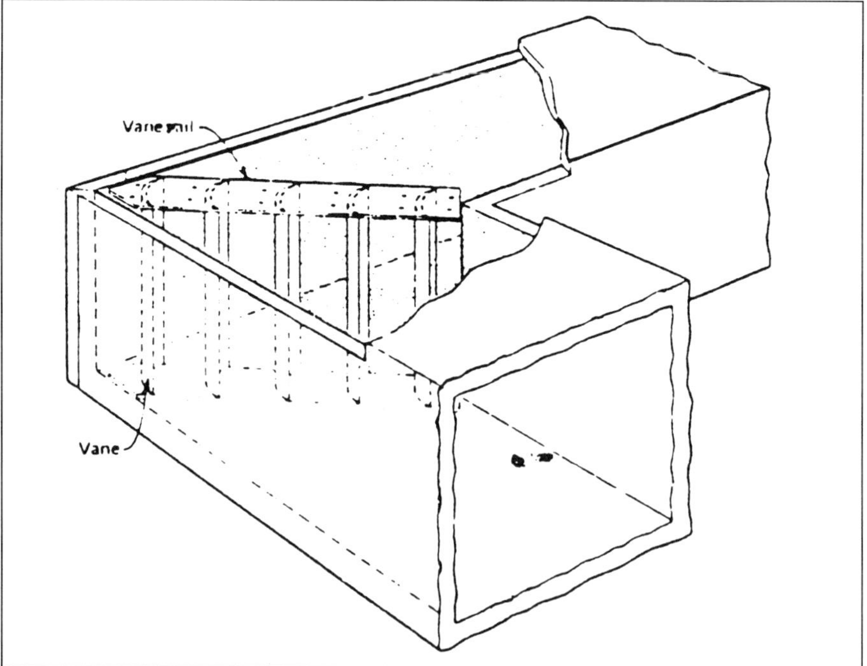

Fig. 3–44 Turning vanes in 90° duct

outside air. If both inside and outside temperatures are sensed, they can be used to derive a control signal that will activate dampers and fans to ventilate the building with outside air. This may be accomplished by an "economizer," which switches the airflow from a normal return air loop to an intake-exhaust flow, making direct use of outside air (see Fig. 3–45 and Fig. 3–46). More accurate control of outside air may be achieved using an enthalpy sensor that determines both outdoor temperatures and humidity.

Another modification to the air-flow system recycles energy in the exhaust system and transfers it into the incoming make-up air stream. This has the effect of preheating (in winter) or precooling (in summer) the air

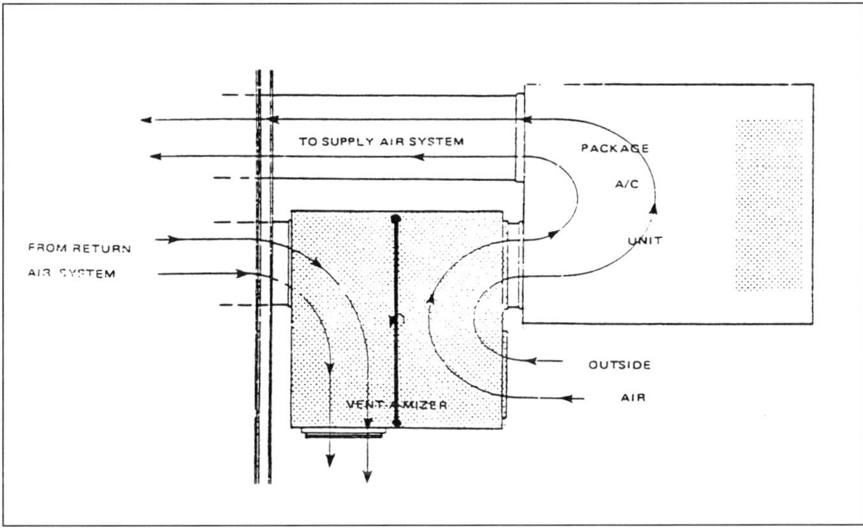

Fig. 3–45 Economizer for rooftop unit

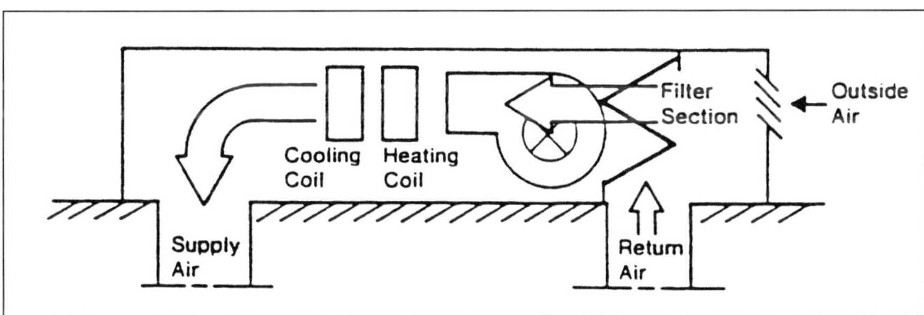

Fig. 3–46 Rooftop unit

entering the building, thus reducing the ventilation load and the primary heating and cooling energy requirements. This is accomplished through the use of a coil loop system, which is comprised of two coils, a pump and control unit, and a liquid-filled loop.

The coils can be looped several hundred feet apart at the existing exhaust and intake ports. The energy used by the pump is negligible compared to the energy recovered. This system is sometimes referred to as the run-around cycle (see Fig. 3–47) and is particularly advantageous where high ventilation rates are required, such as hospitals and some industrial plants.

HVAC Equipment

Controls. These may include thermostats, humidstats, pressure and air velocity sensors, enthalpy controls, time clocks, electromechanical devices, and central control computers. Automatic switching projects should consider local or central control, simple devices vs. complex circuitry, and must be cost-effective. The basic rule applies to "turn it off when not in use."

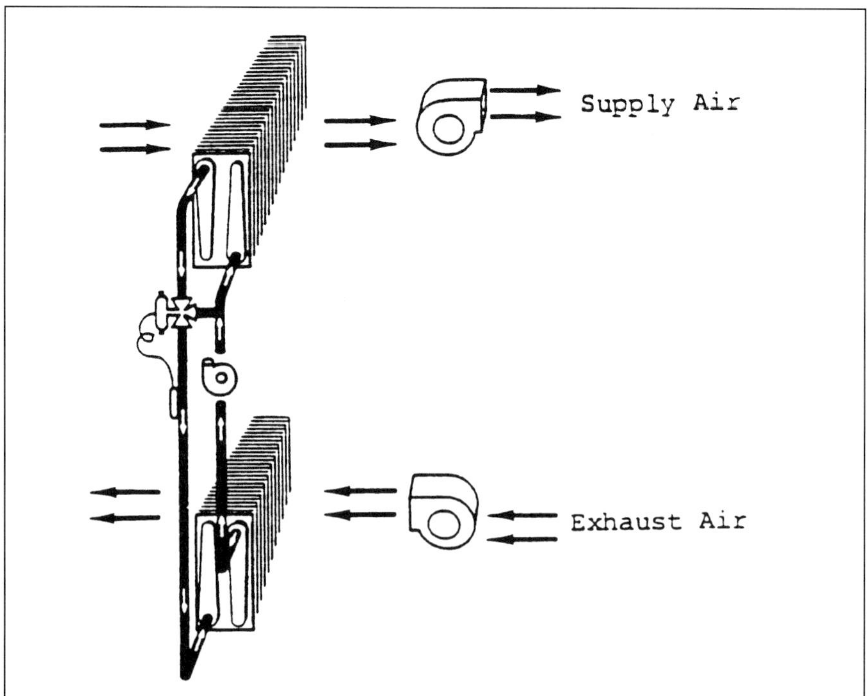

Fig. 3–47 Coil run-around cycle

Air conditioning. Air conditioning may be described as the simultaneous control of temperatures, humidity, movement, and cleanliness of the air in a structure. Prefixes of "summer" and "winter" are added to those systems designed to operate only during summer and winter conditions, respectively. Some systems, designed to operate all year round, are often referred to as heat pumps.

The movement or distribution of air has been described under the heading of ventilation, and cleanliness comprises only the use of suitable filters. For practical purposes, therefore, air conditioning normally used for cooling purposes becomes a matter of refrigeration. Refrigeration essentially removes heat from the air at a lower temperature and rejects it at a higher temperature, expending energy in doing so.

An elementary refrigeration system contains four pieces of equipment (see Fig. 3–48):

- Evaporator or cooling coils
- Compressor
- Condenser
- Expansion valve or throttle that regulates the flow of the liquid coolant

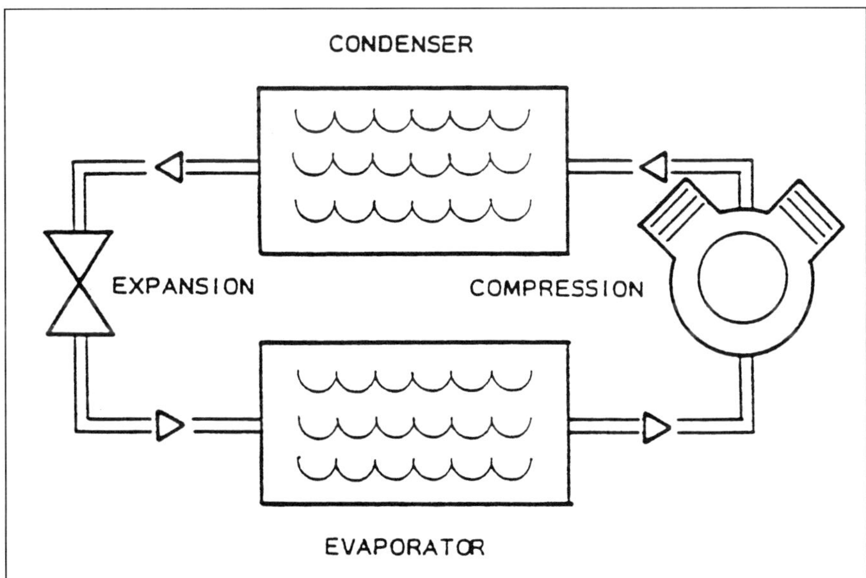

Fig. 3–48 Refrigeration diagram

The cycle of operation begins at the cooling coils, which contain a liquid whose boiling temperature is lower than that of the air surrounding the coils. The heat passing from the ambient air into the liquid coolant causes it to vaporize. The vapor is then drawn into the compressor. There the pressure increases until, at the discharge pressure, the coolant vapor has a temperature high enough that it may be condensed by cooling. The coolant then passes through the expansion valve, emerging as a liquid to the lower pressure of the cooling coils. Fans deliver the cooled air into the ducts for distribution throughout the structure. Fans may also be used to dissipate the heat given off by the condenser.

Control of the refrigerating system is exercised by the action of the thermostat on the expansion valve, allowing it to discharge more or less of the coolant with the cooling coils. For small or residential units, the coolant is usually Freon, while for large commercial and industrial use, the coolant may be ammonia. Air is used as the condensing medium in residential refrigerators. Variable load operating is not achieved by operating the compressor at constant speed and controlling the expansion valve. Instead, it is accomplished with a fixed setting of the expansion valve and on-off operation of the compressor. Large commercial or industrial units may use water circulated by pumps over the condenser coils and cooled through radiators or cooling towers.

It is interesting to note that Freon is being phased out and replaced with compounds having similar characteristics. Freon is a trademark for a group of halogenated hydrocarbons containing one or more fluorine atoms. It is highly evaporative and often leaks from piping and equipment connections, especially when the cooling system is not in use for a prolonged period of time, such as winter.

The gas escapes into the atmosphere, eventually reaching great heights, where it comes into contact with the ozone layer surrounding the earth. It combines with the ozone, creating a concern that it could effectively destroy the protective ozone layer, thus increasing the earth's surface temperature. It is known that the earth does renew the ozone layer. However, it is alleged the destructive rate caused by Freon (and other pollutant gases) is greater than the rate of regeneration by the earth; hence the call for its (gradual) elimination.

Performance of air-cooled condensing units can be improved significantly by precooling the air that passes over the condensing coils.

Evaporative coolers, in which ambient air is passed through a replaceable pad soaked by running water, may be used (see Fig. 3–49). In addition to reducing the temperatures of the cooling air, the pad scrubs out minerals and chemicals that would build up or corrode the cooling coils. The inexpensive pads may be replaced regularly. Improvement in condenser performance is in the nature of 1.1% reduction in energy demand and consumption for each 1° F decrease in condensing air temperature.

Air conditioning with thermal energy storage. For large installations, the system of thermal energy storage may be practical. This system maximizes the duty cycle at a time when condenser efficiency is at a maximum. Condenser efficiency (Btu of cooling per watt-hr) may improve significantly during night hours when temperatures may be 20° F cooler than daytime.

This gain in efficiency may be accomplished by operating the cooling system at night and storing the cooling capacity in the form of chilled water (see Fig. 3–50). Storing cooling capacity at night and retrieving it the next

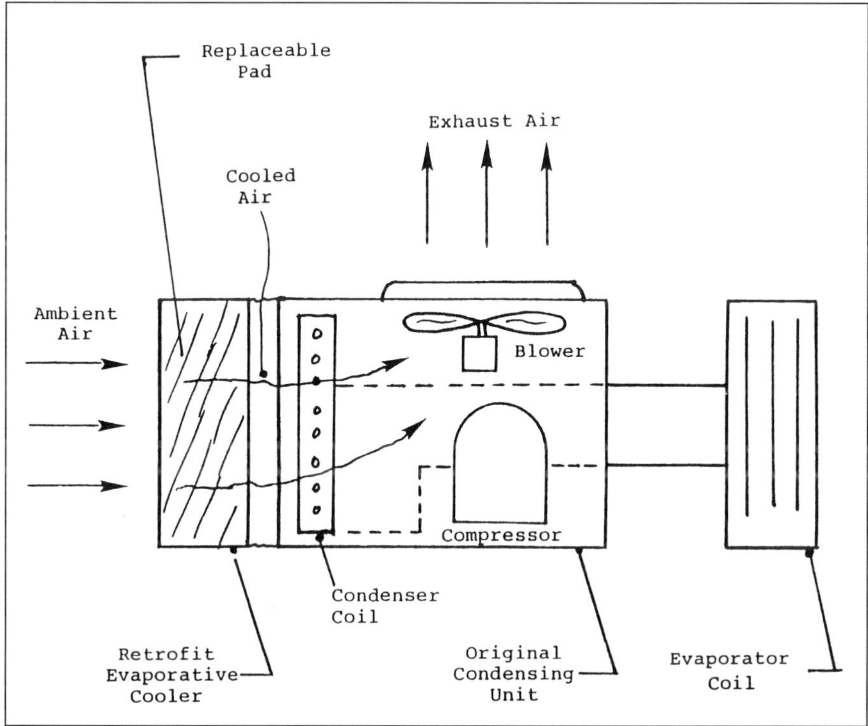

Fig. 3–49 Evaporative boost condenser cooling

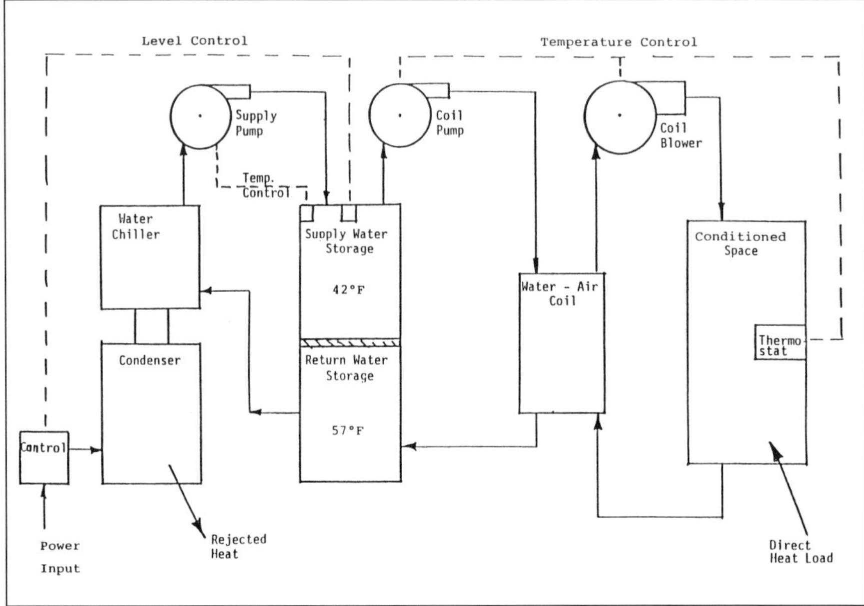

Fig. 3–50 Chilled water storage system diagram

day not only reduces the consumption and demand of electric energy, but shifts the demand to off peak hours. The energy savings with off-peak electric rates can result in substantial reductions in operating costs. Ice also may be used for thermal storage but is not as efficient as chilled water. It does have a volume advantage as a storage medium. Not only is more energy required in its making, but the resultant system is not as flexible as that using chilled water.

Motors. The motors associated with the compressors, pumps, fans, and controls are either single phase for residential equipment and controls or three phase for the larger units, for commercial and industrial applications. Many of the three-phase motors are squirrel-cage induction motors requiring no electrical connection with the rotors.

Other loads. For residential loads, small household appliances (the number, location, and rating) are considered in conjunction with lighting loads. Because of their relatively low power ratings and their diversity factors, they may be only of minor interest and often are ignored.

Refrigerators, hot water heaters, large space heaters, clothes washers and dryers, dishwashers, and ranges are rated as major load appliances, and may substantially affect peak loads or maximum demands. They may be temporarily disconnected for short periods of time so that their loads may be shifted to off-peak periods.

Demand and Consumption Improvement

Suggestions for demand and consumption improvement for individual electrical loads are contained in or suggest themselves in the foregoing description of the various types of loads. More specific suggested improvements are described below. For purposes of discussion, note is taken of the different types of consumers. Six categories are generally recognized as typical by the American Society of Heating, Refrigeration, and Air Conditioning Engineers—ASHRAE. There are flexible lines of demarcation between them:

1. Low-rise residential: dwellings of one or more family units, not more than about three stories high
2. High-rise residential: generally apartments of different sizes and hotels
3. Commercial: office buildings, stores, and small enterprises of many kinds
4. Industrial: manufacturing and servicing plants
5. Institutional: schools, churches, hospitals, and government installations, including military bases
6. Public assembly: theaters, sports arenas, street lighting, and recreational facilities

Loads may be managed in several ways, both to reduce the overall demand and to decrease consumption. These are not listed in order of precedence and may be exercised simultaneously:

1. Reduction in the size of units to serve the load; that is, elimination of oversized, partially loaded equipment
2. Increase in efficiency; that is, replacement of low, steady-state efficient units, even at full load (generally because of low first-cost considerations)
3. Reduction in operating duration time; that is, turning off loads when and where not needed

4. Shift in operating (demand) time; that is, to eliminate or reduce the partial coincidence or contribution to the time of maximum demand of the total installation

Items 1 and 2 are sometimes referred to as retrofitting, while items 3 and 4 are sometimes referred to as peak suppressing. They may be accomplished manually, automatically, or by a combination of the two.

Low-rise Residential Dwelling Consumers

Low-rise residential lighting. Here the demand, as well as consumption, of electricity (energy) may be reduced by replacement of incandescent lamps with fluorescent (gas-filled discharge type) ones. The usual method calls for a change of screw-base type lamp sockets to those of the replacement fluorescent lamp. The cost of such installations must be compared with any savings in operating costs. The fluorescent lamps use less electrical energy and have a longer life than incandescent lamp installations with approximately the same light output. However, they are more expensive to purchase.

There is a screw-base type fluorescent lamp available that fits into most existing screw-base sockets, making the installation of new fluorescent facilities unnecessary. Presently the cost of such lamps is much greater than the equivalent incandescent lamps. It is also more expensive to purchase than the tube-type fluorescent lamp, but the much longer life and the use of existing sockets are factors to be considered in evaluating their economic justification.

Other loads. Other residential major appliances include refrigerators, deep freezers, ranges, dishwashers, clothes washers and dryers, hot water heaters, central heating, ventilating and air conditioning (HVAC) units. Some of these units may be retrofitted profitably.

For example, ranges can replaced (or used less) with microwave ovens or with magnetic field induction top units, which are more efficient than resistance units. Agitation clothes washers can be replaced with allegedly more efficient tumbler action washers. Finally, heating and air conditioning units can be replaced with heat pumps, while other units, where and when warranted, can be replaced with higher efficiency units.

Demand control. Units with relatively high peak demands may be connected to separate circuits, where switching may be controlled by one or more computerized devices. These may be programmed to have the combined demands of the loads they control at any one time not to exceed some predetermined value. Their energization can be set at such times as not to coincide with essential lighting and other preferred loads. For example, dishwashers and clothes washers may be programmed to operate at night and early morning hours, when lighting and television loads are not apt to be turned on. Or the clothes washer can be programmed not to operate at the same time as the dishwasher and refrigerator, so as not to exceed the predetermined overall maximum demand.

These larger units may also be metered separately and billed at special incentive rates. Individual or groups of these units may also be controlled from an external control source employing accepted communication methods. These may include normal telephone lines (also adapted to remote meter reading), separate telephone lines, or fiber optic lines. Other methods might be by power line carrier, microwave, radio (one-way and two-way), cellular, paging devices, satellites, and other electronic devices. A variety of relatively inexpensive timing devices, coordinated in their operation, may also be used. The method chosen should be economically justified and, obviously, greater results may be expected when applied to larger consumers.

Nonelectrical considerations. The color of the surrounding ceilings, walls, floor coverings, and some furniture coverings can enhance the efficiency of room lighting. Floor and portable lamps, placed at strategic points to supplant or enhance ceiling lights, focus lighting in specific areas.

Curtains and drapes not only may enhance room lighting but aid in the heating (in the winter) and cooling (shading from sunlight) in summer. They can act as insulation by maintaining an air cushion between windows and the room ambient air. Other possibilities are insulating ceilings and walls and replacing glass windows and doors with thermal glass panes or installing storm (double-glass) windows and doors. These all act to prevent loss of heat and contribute to lowering demands and consumption of electricity.

High-rise Residential Loads

Lighting. The same discussion concerning low-rise residential lighting loads applies here. Additional considerations are corridor, stairway, foyer, and other stationary lighting as well as lighting for associated indoor automobile parking. All constitute an evening and nighttime load and may employ high-intensity discharge lamps of greater efficiency than incandescent or fluorescent lamps. Retrofitting with fewer higher wattage lamps installed on higher structures and spaced farther apart may result in reduced demand and consumption of electricity.

Other loads. Except for the larger hot water, heating, and air conditioning loads, the same discussion concerning those for low-rise residential loads applies here. Many high-rise residential buildings employ a central laundry. In most cases the clothes washing and drying units are larger versions of the smaller ones, and the discussion concerning low-rise residential loads applies here. Diversity here may not be as great, however, as these units are apt to be used in evenings, after work hours, by tenants. Their peaks may coincide more or less with lighting load peaks, unless some incentive is provided to have them used (for instance) in early morning hours.

HVAC. Hot water and space-heating requirements generally are supplied from other fuels and may be supplemented by electrically operated heat pumps. Where such fuels are not available or are economically unjustifiable (e.g., desert regions), electric resistance heaters may be employed.

Water may be heated to temperatures considerably above utilization values and stored in relatively large containers during off-peak hours. The heated water may be passed through heat exchangers for heating air to be delivered to space heating. The same very hot water may be mixed with colder water to attain the desired utilization temperature to supply hot water needs. In the summer, heat extracted from associated air conditioning units may furnish part or all of the heat requirements for hot water.

Air conditioning requirements may employ the thermal energy storage system, supplemented by heat pump where necessary. Where practical, air conditioning needs may be supplied by two or more smaller units, permitting the shutdown of units not necessary and the fully loading of the one or more necessary ones. A similar arrangement of supplementary heat pumps will also tend to result in reduced demand for electricity.

Elevators. In general, elevator cars and counterweights operate through a system of sheaves or pulleys and cables. Motors are required to overcome the friction of the cables and sheaves and to provide the kinetic energy to start and stop the operation. When the weight of the car exceeds that of the counterweight, the motor supplies the extra energy needed when the car travels up against the force of gravity.

It may be well to have elevator requirements filled by two or more elevators, not only for obviously safety contingencies, but for economy of operation. In a bank of small elevators, some of the units may be taken out of service during nonpeak load periods, allowing the remaining unit's motors to operate more efficiently at or near full load. Elevators in high-rise residential structures experience peak usage early in the morning and again in the early evening. The latter may contribute to the overall peak demand for the structure.

Noncommercial considerations. The discussion concerning low-rise residential loads also applies here.

Diversity. Peak loads in high-rise residential structures occur in the early morning and again (with a greater demand) in the evening hours. Attention, therefore, should be directed not only at individual usage, but also the requirements of the structure as a whole.

Commercial Loads

Lighting. The discussion concerning low-rise and high-rise residential loads applies here. Commercial structures, however, lend themselves to more subdivision of areas controlled by the same switch. Work categories requiring the same levels of illumination should, where practical, be located in the same area and lamp sizes specified accordingly. When high light levels are required for specific tasks, portable lamps may be used to supplement the area lighting.

In large open areas such as showrooms, banks, and some wide-open office arrangements, lighting may be wired in separately controlled groups so that particular sections may be de-energized when not in use. Only those necessary remain energized. Different arrangements may be made to accommodate particular requirements, but the principle of separation into smaller groups remains the same.

There are instances in which for security or other reasons it may be desirable to leave lights on but dimmed. Dimmers of various types are available that essentially lower the voltage applied to the lamps. Care must be taken to limit the lowered voltages to values that will not cause the arcs (in discharge lamps) to be discontinued, or the color of the light output significantly changed.

Sometimes controls are used in combination to take advantage of their unique capabilities. For example, it is not always desirable to turn on lights when natural light is low, particularly in areas not in use. When photocell controls are used to turn on the lights, sometimes time clocks may be placed in the circuit or part of the unit to keep lights off when the illuminated areas are not in use.

Diversity. Lighting loads for commercial category loads, however, differ from high-rise residential loads in that their peak demands occur at different times. Residential loads peak in the evening, while the commercial load demands generally peak during business hours. Where large-scale advertising or beautification lighting may exist, sharp evening peaks may develop.

The operation of fluorescent lamps at frequencies greater than the usual 60 cycle per second yields increased light output, that is, greater efficiency. The high frequency voltage for such operation of the lamps may be generated by motor generator sets, rotary converters, or electronic inverters. Such frequency measures, especially when central in a distribution system, are obviously very expensive where retrofitting is concerned. However, electronic ballasts are available that provide high frequency outputs while operating on standard power system frequencies. Retrofitting of ballasts may be more readily economically justified, providing they are compatible with those they replace.

For outdoor parking areas, electrical consumption may be reduced by fewer, more efficient mercury vapor lamps, metal halide lamps, or sodium vapor lamps, if the resulting light distribution pattern is acceptable.

HVAC. Heating, ventilating, and air conditioning loads are very similar to those for high-rise residential structures, as previously discussed. The main difference is in the timing and duration of peak loads. Whereas peak commercial loads occur only during the work week (excluding holidays), peak residential loads occur daily.

Electric lighting represents one of the largest internal loads in commercial and industrial buildings. Measures that reduce lighting energy requirements will also reduce the heat load on air conditioning systems.

Example 3.2. Consider a cooling system with an energy efficient ratio (*EER*) of 8.0 (medium relative efficiency). For every 1,000 watts reduction in lighting loads, additional savings in air conditioning energy will be:

$$\text{Savings} = \frac{\text{Reduced heat load}}{EER} \tag{3.16}$$

Using the data given:

$$\text{Savings} = \frac{(1{,}000\ \text{W})\ (3.413\ \text{Btu/hr/W})}{8.0\ \text{Btu/hr/W}} = 427\ \text{W}$$

Total savings from reduced lighting load will be 1,427 watts (the air conditioning saving added to the lighting saving). The air conditioning savings add nearly 43%, a significant additional benefit. It may also be noted that the less efficient the air conditioner, the greater the savings. In this example, if the *EER* was 6.0 (a low relative efficiency), the savings above would be 569 watts, an additional benefit of nearly 57% from lighting load reduction.

Nonelectrical considerations. The same discussion concerning low-rise residential loads applies here.

Industrial Loads

The kinds of industry and their electrical requirements are so varied that generalizations are meaningless. On one extreme are the highly sophisticated manufacturers, such as those associated with computer chips, in which temperature, humidity, and ambient cleanliness call for extraordinary measures. On the other end are the large, open-area, high-ceiling construction and repair shops, where both lighting and HVAC needs may be tailored to individual tasks. Here portions of the discussions for residential loads and, in particular, the discussions of commercial loads could apply.

Where such plants operate continuously around the clock, peak loads may exist at all hours of the day. The reduction of overall demand may be substantial and, hence, worthy of intense examination, including the use of portable lights, fans, tools, and equipment suitable to the task at hand.

Where motor loads may affect the power factor of the electric supply, demand may be reduced through the use of remedial capacitors either at individual motors or on the circuits supplying them (see Fig. 3–51a and Fig. 3–51b). In this instance, provision may be made to switch off all or some of the units that may overcompensate for the correction. Use of synchronous motors for both work and power factor correction should be considered.

Where plants utilize three-phase power supply, balancing of loads between phases (as much as practical) will reduce losses. Losses may be reduced in the neutral conductors (in Y circuits) or in circulating currents (in Δ circuits), affecting the overall demand and consumption of electrical energy.

Air conditioning. Because air conditioning may be a significant part of energy consumption and operating costs, it may be well to consider the replacement (retrofitting) of a less efficient system with one of greater efficiency. This is especially beneficial if maintenance costs are relatively high.

Example 3.3. A factory has a 50-ton, direct expansion (DX) rooftop cooling unit, having an energy efficiency ratio (*EER*) of 5.0 and operating approximately 1,500 hours per year. Its maintenance results in $900 in labor charges annually. Management is considering replacement with a more efficient unit of 50 tons having an *EER* of 9.0 and estimated annual maintenance of only about $200. Installed cost of the new system (condenser replacement only) is estimated at $25,000. The replacement equivalent to the present system is estimated at $15,000. Hence, the "incremental cost" associated with any energy savings would be only $10,000. Calculate the breakeven for the equivalent replacement system versus the upgraded system.

Annual energy consumption of the present system is calculated using Eq. 3.14:

$$kWh_{present} = \left(\frac{(50 \text{ tons})(12{,}000 \text{ Btu/hr/ton})}{(5.0 \text{ Btu/hr/watt})(1{,}000)}\right) \times 1{,}500 \text{ hr/yr} \qquad (3.17)$$

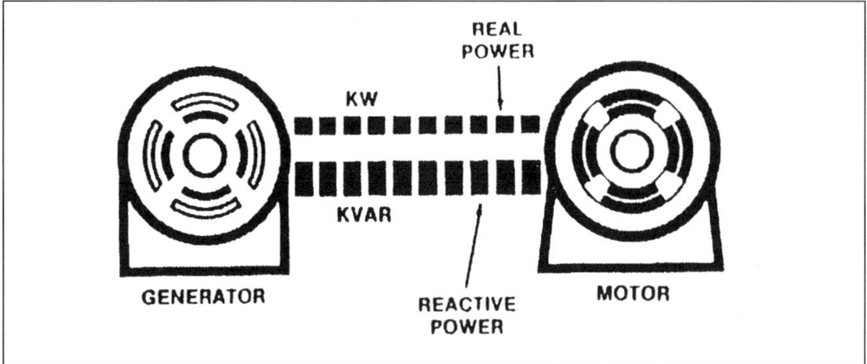

Fig. 3–51a An induction motor operating under partially loaded conditions without power factor correction (The feeder line must supply BOTH magnetizing and useful currents.)

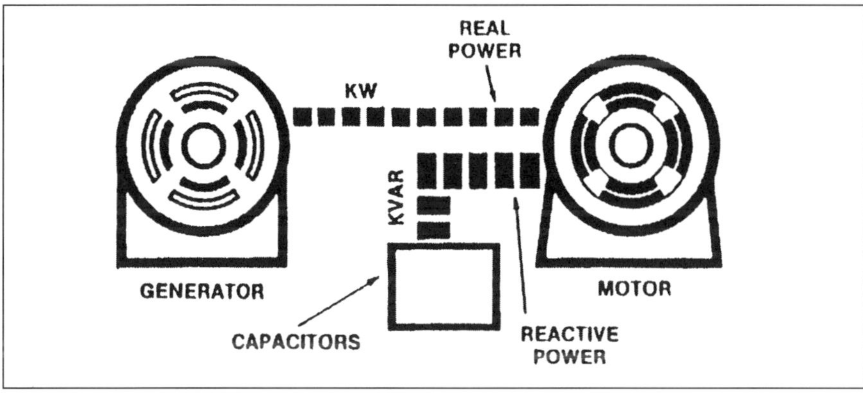

Fig. 3–51b The result of installing a capacitor near the same motor to supply the magnetizing current required to operate it. The total requirement has been reduced to the value of the useful current only, thus either reducing power cost or permitting the use of more electrical equipment on the same circuit.

$$kWh_{present} = 180{,}000 \text{ kWh/yr}$$

Annual energy saving with new unit:

$$kWh_{savings} = kWh_{present} \times \left(1 - \frac{EER_{present}}{EER_{new}}\right)$$

$$kWh_{savings} = 180{,}000 \text{ kWh/yr} \times \left(1 - (5/9)\right)$$

$$kWh_{savings} = 80{,}000 \text{ kWh/yr}$$

Estimated energy cost is 6¢/kWh (block rate). Consequently, the estimated annual cost saving is:

$$\text{Annual savings} = 80{,}000 \text{ kWh} \times \$0.06/\text{kWh} = \$4{,}800/\text{yr}$$

Based on the calculated annual savings, the breakeven periods for an upgraded replacement versus a standard replacement can also be calculated (neglecting any interest incurred on loans needed for the purchase):

$$\text{Breakeven} = \frac{\text{Cost of equipment}}{\text{Energy savings} + \text{Maintenance savings}} \tag{3.18}$$

Substituting the data for equivalent and upgraded replacements:

$$\text{Replacement Breakeven} = \frac{\$15{,}000}{(\$0) + (\$900/\text{yr} - \$200/\text{yr})} = 21.4 \text{ years}$$

$$\text{Upgrade Breakeven} = \frac{\$25{,}000}{(\$4{,}800/\text{yr}) + (\$900/\text{yr} - \$200/\text{yr})} = 4.48 \text{ years}$$

Institutional Loads

The discussion on lighting and HVAC requirements for residential and commercial loads also applies here. The major differences pertain to the time and duration of peak demands. For schools, demands will occur in the morning and afternoon hours, five days weekly, excluding holidays, and endure about 10 months of the year. The exclusion of summertime may affect air conditioning loads substantially. Special events may introduce occasional evening peak demands and may generally be neglected. Churches generally experience peak demands on Sunday mornings, with other lesser demands occurring with special events (weddings, funerals, holy season events, etc.). Located usually in residential areas, these peak demands have little effect on electrical demands for those areas.

Hospitals, on the other hand, have large and continuous lighting and HVAC requirements. Here, waste heat from air conditioners may be used to meet the demands for heat and hot water. Electrical operations are continuous, with extraordinary peaks experienced when emergencies arise (accidents, epidemics, etc.).

Government installations may run the greatest of demands:

- Prisons requiring varying demands during the day, continuously throughout the year
- Court rooms, museums, and similar activities approximate commercial loads
- Police, fire, water supply, airports, lighthouses, port facilities and military bases may be considered 24-hour daily loads, with relatively small variations in demands and consumption

Public Assembly

Here, too, the discussions associated with residential and commercial loads apply. As with institutional loads, demands vary as to the time of occurrence and duration. Theaters may require relatively large heating and air conditioning loads during periods of operation that may vary from regular daily showings to less frequent periodic showings. Occasional special events may introduce relatively large demands and consumption of electrical energy.

Recreational facilities may have irregular lighting loads depending on their character. Parks, however, have loads approximating those of street lighting: full load demand during the period they operate, usually during the night hours.

Some reduction in street lighting consumption may be realized by individual control of lamps by photocells mounted on the lamp or supporting structure. Demands occur during off-peak hours and have little effect on system peak demands. However, more efficient lighting units and patterns may contribute to lessening of losses (and investment expenditures).

Implementation

Many of the suggested improvements in demands and conservation of electrical energy as a result of load management can be justified on their own merits. Where economic justifications may not be realized, advantage should be taken of incorporating the suggested changes with maintenance programs (particularly preventive maintenance programs), revamping or expansion programs, and other construction activities. The incorporation of other changes may justify economically some proposed load management changes.

Some methods for determining economic justification of proposals are given in Appendix B, "Economic Studies." Additional information concerning electrical energy is found in Tables 3–6, 3–7, 3–8, and 3–9.

Table 3–6 Convenient References

Energy Conversion Factors		
Multiply	**By**	**To Obtain**
Watts	3.413	Btu/hr
Kilowatts (kW)	3.413	Btu/hr
Kilojoules (kJ)	0.9478	Btu
Therms	100,000	Btu
Horsepower	2,575	Btu/hr
Boiler Horsepower	33,472	Btu/hr
Horsepower	0.7457	kilowatts (kW)
Horsepower	550	ft-lb/sec
Btus/hour	0.000293	kilowatts (kW)
Btu/hour	0.293	watts
Tons (refrigeration)	12,000	Btu/hr

Table 3–7 Abbreviations

KWh	Kilowatt-hours
Btu	British thermal unit
Mbtu	1,000 Btu
MMBtu	10^6 Btu
CCF	100 cubic feet of gas
MCF	1,000 cubic feet of gas
PSIA	Pounds per square inch, absolute pressure
PSIG	Pounds per square inch, gauge pressure
BBL	U.S. barrel = 42 gallons

Table 3–8 Fuel energy content

Fuel Type	Energy Content/Unit
Natural Gas	1.03 × 106 Btu/MCF
#2 Diesel Oil	5.83 × 106 Btu/BBL
#6 Residual Oil	6.29 × 106 Btu/BBL
Propane	3.84 × 106 Btu/BBL
Butane	4.28 × 106 Btu/BBL
Gasoline	5.25 × 106 Btu/BBL

Table 3-9 Power Formula for Electric Motors

Voltage and current known:

$$kW = \div \Theta \times \text{volts} \times \text{amps} \times \text{P.F.}$$

where:

$\Theta = 1$ for single-phase motors
$\Theta = 3$ for three-phase motors

Horsepower rating known:

$$kW = \frac{\text{HP Rating} \times 0.746 \times \text{L.F}}{\text{M.E}}$$

where:

P.F. = Power factor (cosine V – I lag angle)

$$\text{L.F.} = \text{Load factor} = \frac{\text{actual load}}{\text{rated load}}$$

M.E. = Motor efficiency at actual load

HVAC Terminology

There are many terms, abbreviations, and acronyms associated with heating, ventilation, and air conditioning (HVAC) systems. Definitions of those most commonly used in the analysis, design, operation, and maintenance of such systems may prove helpful.

AHU. Air-handling unit; contains a blower, filters, and heat exchangers for distribution of air to a specific area, building, zone, or zones.

Air change. A measure of the ventilation rate for a given space, usually expressed as air changes per hour.

ARI. Air Conditioning and Refrigeration Institute.

AGA. American Gas Association.

ASHRAE. American Society of Heating, Refrigeration, and Air Conditioning Engineers.

Boiler. A basic heating system used to generate steam or hot water, used with hydronic systems.

Boiler hp. A unit of capacity in steam boilers, equal to 33.472 Btu/hr.

Btu. British thermal unit, a basic unit of heat energy; the heat required to raise 1 pound of water 1° F.

Building envelope. The exterior surfaces of a building; the walls, ceiling, and floor.

Cfm. Cubic feet per minute; flow rates adjusted for standard temperature and pressure may sometimes be referred to as scfm.

Chiller. A refrigeration machine used to chill water in a hydronic HVAC system.

Condenser. The heat exchanger in a refrigeration system that rejects heat as it converts the high-pressure refrigerant from a gas to a liquid; either air-cooled or water-cooled.

COP. Coefficient of performance; the ratio of delivered Btu (heating) or absorbed (cooling) to the Btu input.

Degree days. A measure of the severity of the weather; the annual sum of the number of degrees (usually Fahrenheit) of each day's mean temperature above (cooling) or below (heating) a reference temperature, normally a 65° F base.

DX. Direct expansion; the type of basic cooling cycle used generally in larger cooling systems.

Economizer. An arrangement of ducts, dampers, and controls that permits the introduction of outside air into air-handling (heating and cooling) systems.

EEI. Edison Electric Institute.

EER. Energy efficiency ratio for air conditioning systems, usually expressed in Btu/W-hr (*see also* **SEER**).

EMS. An energy management control system; usually an electronic system that can be used to switch or modulate HVAC system functions and to control other energy-consuming devices (*also* **EMCS**).

Evaporator. The heat exchanger in a refrigeration system that absorbs heat as it converts the low-pressure refrigerant from a liquid to a vapor.

Evaporator boost cooler. An auxiliary device that lowers the condenser air temperature, thereby raising the effective energy efficiency ratio (*EER*) (*also* **Precooler**).

Heat pump. A reversible refrigeration machine that can provide either heating or cooling.

Infiltration. The leakage of air into a building through space or cracks around doors, windows, etc.

kW. Kilowatt; equal to 1,000 watts; a unit of measure of electric power.

kWh. Kilowatt-hour; a unit of electric energy equal to a kilowatt load operating consuming energy) for 1 hour.

M & O. Maintenance and operating procedure.

Precooler. *See* **Evaporator boost cooler.**

R-value. The thermal resistance of an insulating material, expressed in hour/ft^2/temperature (in ° F)/Btu.

SEER. Seasonal energy efficiency ratio; for small air conditioning systems reflecting the effects of seasonal partial loads and cycling; usually expressed in Btu/watt-hour (*see also* ***EER***).

TES. Thermal energy storage.

Thermal load. The heat energy that must be removed (cooling) or supplied (heating) by HVAC systems, usually expressed as tons or Btu/hr.

TSE. Thermal system efficiency; the ratio of thermal output to thermal input energy.

Ton. A unit of capacity in air conditioning systems; equal to a cooling rate of 12,000 Btu/hr.

U value. The transmission coefficient (thermal conductance) of an insulating material, expressed in Btu/hr/ft^2/ ° F.

VAV. Variable air volume; an energy-saving design for air distribution systems.

Zone. An area within a building under temperature control of one specific thermostat.

4

Utility Control

Utility systems have both direct and indirect control of actions leading to the end result of reducing the total load on the system. The direct control of actions is generally from a central location of system operations and/or a divisional distribution operation center. Indirect control takes the form of design parameters, design standards, marketing programs, rate structures, and conservation programs produced by utility staff services.

Direct Action

Electric load management and distribution automation (DA) are interrelated, complementing each other, but they have different objectives. Electric load management is primarily involved with reduction of demand and energy. DA is primarily involved with remote control of distribution system devices to improve service reliability. It includes computerization of support services such as interruption management and distribution system component location. DA may also include automatic operation of devices at a remote location acting with intelligence of system conditions.

A good example of combining electric load management and DA would be the use of voltage control to temporarily reduce a system demand. From a central control center, an operator can activate a voltage reduction device in a remote substation to automatically drop the outgoing distribution bus voltage. This is usually by increments of 3% and 5%, according to system conditions. Lowering the voltage to consumer loads will produce an immediate drop in demand, particularly with resistive-type loads. Generally, a utility will conduct a test of predicted results under favorable system conditions, using the actual results to estimate load drop available for future system contingencies or for regular peak shaving. Each system will react differently depending on the composition of its consumer loads.

Another example of direct utility control would be to temporarily interrupt service to consumer loads in such a way that the immediate effect would not be noticed by the consumer. The load would be offset to another time period, sometimes to a time completely different than the system peak. Electric water heaters, swimming pool pumps, and air conditioners are typical of such consumer summer loads. Figure 4–1 shows the radio antenna and control interface designed to receive a signal from a central operations location that interrupts the power supply to a large swimming pool circulation pump. At the utility operator's discretion, or by a programmed computer action, the swimming pool pump is turned off for a required amount of time to reduce its contribution to the system peak demand.

The direct control of this one device would have virtually no effect on a system peak, but when multiplied by thousands in a large system, its effect would be in megawatts. The pool pump (or water heater or air conditioner) can be cycled on and off during peak load periods. This can also be done in the event of a system emergency of extremely higher than expected load or a generator outage requiring reduction of load rapidly. In these instances, the utility has direct control technology that is useful for electric load management purposes. It also contributes to system reliability under emergency conditions, perhaps avoiding a complete system shutdown.

Another advantage of the DA load control devices is that they can be programmed to prevent problems in cold load pickup after an extended outage. This is done by staggering the time the individual loads come back after the area is reenergized. Without this feature, if there were 1,500 individual 2-kW loads interrupted, when the area is reenergized, they would all tend to come on simultaneously at full load, or in this case, 3,000 kW. This load pickup in addition to other loads may cause overcurrent relays to operate and deenergize the circuit again, requiring sectionalizing and restoring load in segments.

Fig. 4–1 Radio control of pool pump

Indirect Design Control

One of the most important considerations in designing and operating an electric system is to control the magnitude and cost of losses in real and reactive power and energy. In general, losses can be reduced by the utility design of components and the coordinated operation of the whole system. This should occur with an economic evaluation of the incremental additional cost to effect the reduction. Losses on a system contribute to the peak load and capacity on each element of the system must be provided for as well as the consumer loads. Monetary losses can be calculated by multiplying energy losses in kilowatt-hours by the cost of producing this energy. This allows the economic evaluation of increased investment to reduce losses versus the cost of the losses. There are many variables that enter into an economic evaluation. These include load factor, anticipated rate of growth, inflation rate, capital investment costs, labor, maintenance of operation costs, incremental cost of energy, type of system, and taxes.

Losses occur in transformer and line conductors of the transmission, substation, and distribution system. The majority of losses (60%) are in the distribution system for an average system. Table 4–1 gives the allocation of losses in a typical system.

Table 4–1 Losses in a Typical System

	% Loss As a % of Load		Load Factor
	Power Loss	Energy Loss	
Generation Step-Up Transformer	0.7	0.8	50
Transmission	4.4	2.5	45
Substation	1.1	1.3	40
Primary	3.5	1.7	38
Distribution Transformer	1.7	2.2	36
Secondary	3.1	1.6	34
Service	1.2	0.4	15

Source: Ebasco Services

Transformer loss control

Transformer loss can be controlled through parameters assigned by the utility to the manufacturing design. This design should take into account the incremental energy cost of production, the cost of core losses, and the cost of copper losses. These design parameters can be used to arrive at the most efficient unit purchased for minimum losses and an economic operating and capital cost.

Reactive Loss Control

In addition to real power and energy losses on a system, there are losses incurred from supplying reactive components. On the utility system itself, lagging reactive losses result from step-up transformers, step-down transformers at substations, and distribution transformers connected to the primary lines. The other major source of lagging reactive losses comes from consumer load devices—motors used in refrigeration, air conditioning, pumps, elevator motors, nonincandescent lighting, as well as other industrial processes.

The ideal system would operate at unity power factor to minimize reactive losses. There are some "built-in" loading reactive sources such as transmission cables. However, the principal device used to offset lagging reactive losses and bring a system as close as possible to unity power is the capacitor. The capacitor may be installed in banks at substations, but by far the most common use is on the primary distribution lines, where they are installed for voltage regulation purposes. Capacitors will result in voltage rise to offset normal voltage drop by resistive and lagging reactive loads on a circuit. Distribution capacitor banks may be installed connected to the primary line at all times (a fixed bank—see Fig. 4–2). It can also be installed as a switchable bank (see Fig. 4–3) used both for voltage regulation and loss control.

Conductor Loss Control

The control of losses from conductors used for transmission and distribution (primary and secondary) lies in the economical sizing of the conductor for a particular installation or design standard. Instead of selecting a conductor size to match the maximum load expected, greater size conductors would be evaluated to compare additional capital cost to savings resulting from reduced power and energy losses.

Another significant element in conductor loss control is the economic dispatch of generation coupled with conductor losses. In this utility control, the cost of losses in transmission is combined with the cost of genera-

Fig. 4–2 Fixed capacitor bank

Fig. 4–3 Switched capacitor bank

tion through system load flow simulation studies. The results of the studies would indicate the least cost combining generation and line loss costs to produce a recommended operating scenario to the system operator.

Indirect Control of Consumer Power and Energy

The utility may implement programs to influence consumer decisions in energy-saving devices and actions by the consumer that require a capital expenditure. These programs are encouraged by regulatory agencies as well as environmental groups. Some aspects of such programs may also require additional costs by the utility. These costs then have to be evaluated for economic feasibility, comparing savings from the results to the cost of implementing and maintaining the program.

Time of Day and Seasonal Electric Rates

With the approval of regulatory authority, a utility may establish a price of electricity schedule designed to reduce peak loads both on a daily and seasonal basis. Figure 4–4 shows a price structure designed to reduce a summer peak by charging a very high price for usage during the peak hour of the summer period. There are varying prices for other times of day and seasons. Table 4–2 shows typical appliance loads, their usage, and the effect of a consumer electing to avoid high-price usage to the extent possible or preferable in terms of comfort and convenience. This can be achieved by the consumer manually controlling usage or installing automatic load-controlling devices. Remote meter reading systems can also be used, at the consumer's election, not only to schedule usage but also to display the current energy cost to the consumer.

This is relatively simple for residential consumers, but becomes more complex for commercial and industrial applications. As an example, a typical building electric demand profile is shown in Figure 4–5 before adjusting loads to lower the peak demand. Engineering personnel would review production schedules, individual work processes, and work shifts to determine what loads could be moved to off-peak hours. This perhaps requires more planning and management than direct capital investment and is usually implemented through automation methods rather than changing equipment or depending on manual control. When implemented, the load profile may result in lower peak demand, but the same energy consumption, as shown by Figure 4–6.

Table 4-2 Appliance Usage and Cost

Appliances	Total	Low price	Moderate price	High price	Critical price	Average daily cost
Central heat and air conditioner	$74.52 Credit $7.32 =$67.20	$33.35	$23.21	$5.95	$12.01	$2.40
	854kWh	600kWh	213kWh	17kWh	24kWh	27kWh
Water heater	$21.21 Credit $1.00 =$20.21	$11.95	$9.28	$0	$0	$0.68
	375kWh	290kWh	85kWh	OkWh	OkWh	12kWh
Pool pump	$24.48	$19.90	$4.58	$0	$0	$0.79
	400kwh	358kWh	42kWh	okWh	OkWh	l2kWh
Clothes dryer	$8.34	$8.34	$0	$0	$0	$0.27
	150kWh	150kWh	OkWh	OkWh	OkWh	4kWh
Other	$64.02	$29.23	$9.48	$6.30	$19.01	$2.07
	594kWh	451kWh	87kWh	18kWh	38kWh	19kWh
Customer charge	$7.04					
Monthly total	$191.29	$102.77	$46.53	$12.25	$31.02	$6.21
	2373kWh	1849kWh	427kWh	35kWh	62kWh	76kWh

Source: Central Power and Light: Laredo. Texas

Consumer load profiles are an integral part of the electric load management program. Through them the utility and the consumer can exercise control of peak usage and the use of energy at lower costs by scheduling loads. In the past, consumer load profiles were done with portable recording metering installations on a statistical sampling basis. Technology is now available to provide such profiles to the utility and the consumer by the use of automatic remote meter reading devices, communication systems, and computer programming.

An example of this is the American Innovations system. It uses a remote meter reading device installed under the cover of a conventional electric meter, connecting it to the consumer's telephone line, and transmitting the data back to a central receiving point. There a computer can store the

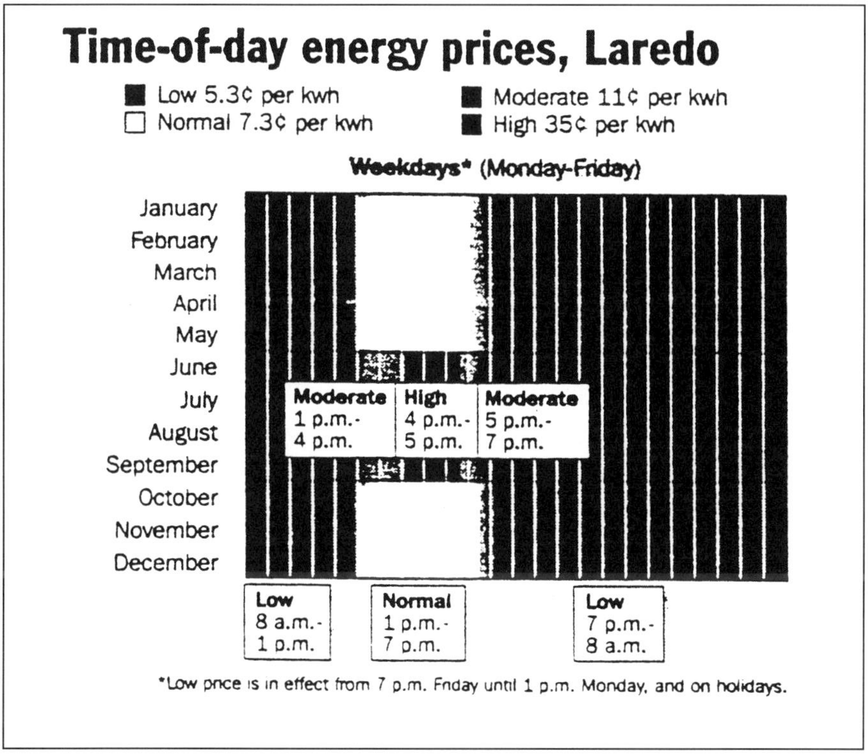

Time-of-day energy prices, Laredo
Low 5.3¢ per kwh
Normal 7.3¢ per kwh
Moderate 11¢ per kwh
High 35¢ per kwh
Weekdays* (Monday-Friday)
January
February
March
April
May
June
July
August
September
October
November
December
Moderate 1 p.m.- 4 p.m.
High 4 p.m.- 5 p.m.
Moderate 5 p.m.- 7 p.m.
Low 8 a.m.- 1 p.m.
Normal 1 p.m.- 7 p.m.
Low 7 p.m.- 8 a.m.
*Low price is in effect from 7 p.m. Friday until 1 p.m. Monday, and on holidays.

Fig. 4–4 Peak load pricing (*courtesy Central Power and Light, Laredo, TX*)

KW DEMAND
1600
1400
1200
1000
800
600
400
200
12 1 2 3 4 5 6 7 8 9 10 11 12 1 2 3 4 5 6 7 8 9 10 11 12
AM
PM

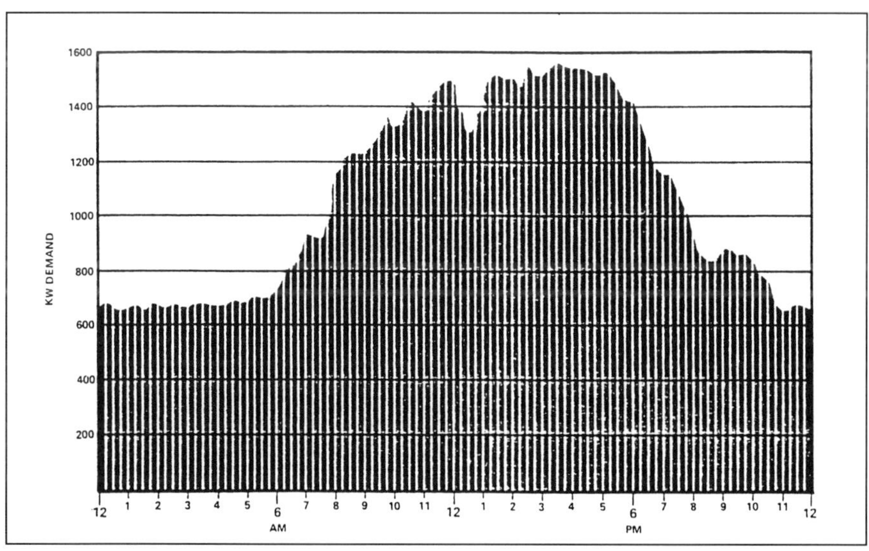

Fig. 4–5 Building load profile

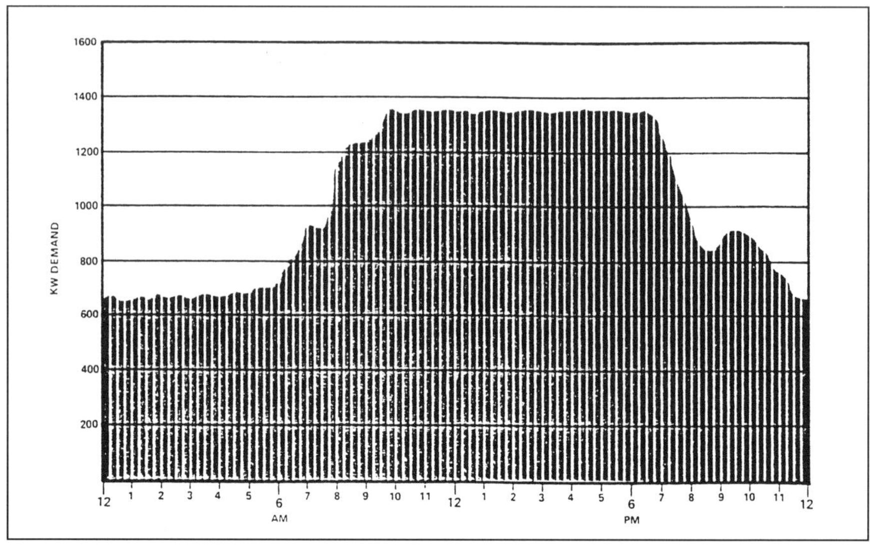

Fig. 4–6 Modified building profile

information, calculate the consumer's bill, and develop load profiles of the usage. It has the capability, for instance, to record data hourly (or in shorter periods). This may be important if future rate schedules are aimed at hourly purchases at different prices or purchasing energy from competitive sources at different times of the day.

Conservation

Utilities can approach conservation in many ways:

- Furnishing free energy audits
- Furnishing technical assistance in initial design or modifications of buildings
- Offering incentives for the consumer to use more efficient devices
- Advertising alternatives and sources of conservation assistance
- Using bill inserts and billing design to call attention to usage and cost

Conservation programs are many and varied according to the geographic region, climate, consumer classifications, and utility configuration. However, all are aimed at reducing energy consumption and system peak loads. Examples of programs are shown by Figure 4–7 and Figure 4–8.

Some bright news about lower electric bills.

Did you know that lighting can account for up to 15 percent of your electric bill? But there's a way you can reduce this cost by as much as 75 percent — without sacrificing the quality of your lighting.

Today's compact fluorescent bulbs consume one-fourth the energy of a standard incandescent light bulb and last up to 10 times longer, so each bulb could save you up to $100 over its lifetime.

What's more, LILCO is currently offering a $6 rebate on each qualifying fluorescent bulb you purchase until December 31, 1997. To take advantage, simply pick up a LILCO rebate brochure at local home improvement stores or call our Energy Hotline at 1-800-692-2626.

So make the switch to fluorescent lightbulbs. You'll lower your electric bill while conserving energy.

And that's bright news for everybody.

Fig. 4–7 Lighting incentive advertisement (*courtesy LILCO*)

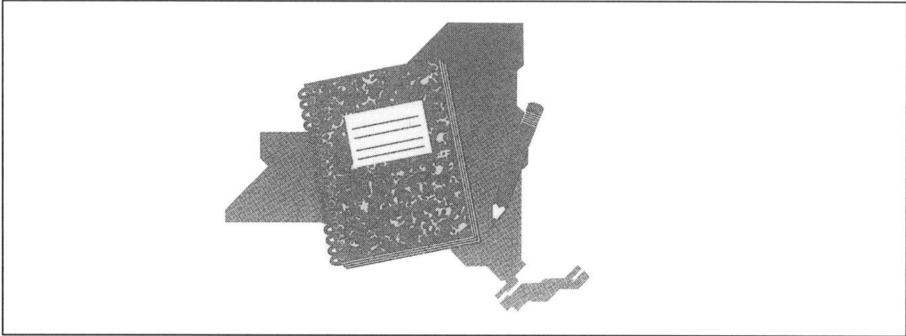

Fig. 4–8 Assistance program (*courtesy New York Power Authority*)

Other elements leading to reduced system demand

Consumer self-generation in the form of photovoltaic panels, small wind generators, small gas-fueled electric generators, and even fuel cells is creating the possibility of a future impact on utility system peaks. This distributed generation will probably be more significant for commercial and industrial application than for residential applications. This is due to minimum sizes needed for economic use and the likelihood of restrictions on noise and pollution in residential areas.

5

Energy Pricing and Demand

Many factors enter into the pricing of a commodity, and this applies to electric service, referred to generally as rates. Ideally, rates should reflect cost of service and an equal rate of return for all categories of consumers. They should be simple to understand and relatively inexpensive to administer. Load management attempts to maintain these criteria in tailoring special rates to foster the goals of lesser demand and reduced consumption of electrical energy.

A basic electric utility rate includes the following, measured and billed on a periodic cycle, usually monthly:

- Consumer service charge
- Energy charge based on actual usage
- Current fuel adjustment charge
- Demand charge—based on the highest demand during the particular billing period

There are other features included in rate schedules designed to encourage and compensate or penalize consumers in order to meet the goals of load management. These include:

- A deviation from normal consumption rates when demand exceeds a predetermined value (block)
- Low power factor
- Peak vs. off-peak pricing (times of day and time of year)
- A demand "ratchet" clause that takes the highest monthly demand as a base for new minimum demands for future billing periods

Two other items that may also appear on consumer billings, although not included in rate schedules, are:

- Cogeneration power cost factor
- Taxes—sales and other types

Consumer Service Charge

The consumer service charge is an "overhead" charge, independent of electrical consumption and demand. In addition to costs of metering, meter reading, and billing, it includes some costs of management, engineering, stores, insurance and claims, building maintenance, and other consumer service items.

Energy Charge

The energy charge covers the major portion of the costs for establishing, operating, and maintaining sources of energy (power plants, fuel pipe lines, etc.) and transmission and distribution facilities. In many cases, this part of the rate schedule is structured so that there are decreasing charges for greater consumption. This is based on the principle that these costs are generally fixed and spread out over increasing levels of usage, so the cost per unit decreases. (In other enterprises, this entails wholesale and retail prices; cost to the supplier is less for higher volumes, reflected in lower wholesale prices.)

For load management purposes, however, this schedule may be reversed somewhat. The greater usage could imply the use of less-efficient generating units to supply the consumer energy needs, coupled with greater losses in its delivery. Hence, some schedules may call for higher unit costs for the larger blocks of energy delivered, an "inverted" rate schedule.

Current Fuel Adjustment Charge

The current fuel adjustment charge reflects the actual cost of fuel used in the generation of electricity. It is not a fixed charge but, at a given time, it tends to be the same for every unit of electricity generated (in kWh). The fuel rate is established by dividing the total fuel cost by the total electrical energy generated in a given billing period (¢/kWh). This charge for that period becomes a flat rate and is applied to every consumer regardless of consumption levels. This charge may sometimes be a negative one, resulting in a credit or refund to consumers when, for whatever reason, the charge billed is greater than the actual cost.

Demand Charge

The demand charge represents the cost of the capital investment made to meet the peak or maximum demands of the consumer. This charge is based on the highest consumption rate experienced by the consumer during a billing period. This consumption rate is based on the consumer's average consumption use during the highest demand interval, which usually is 15 minutes, 30 minutes, or 60 minutes. Demand is expressed in kilowatts and is equal to the kilowatt-hours used in the interval divided by the time of the interval (expressed as 0.25, 0.50, or 1.0 hours).

The demand charge applies even if the consumer experiences this maximum use of the utility's facilities for only a very small amount of time during the billing period (usually a month). Meters measure kilowatt-hours consumed and the kilowatt demand by the consumer. (The detailed operations involved in the metering are not pertinent to this discussion and are not included here.)

Demand and Consumption

Some rate schedules provide a penalty when the consumer exceeds a predetermined contracted demand. The penalty, however, is charged against consumer's consumption of electric energy. These are referred to as "block extender" rates and generally apply to the larger commercial and industrial consumers. When the demand limit is exceeded, the cheaper (third) consumption block rate is canceled. Instead, the higher (second block) consumption rate applies; the higher the peak demand, the longer the higher "block extension" rate is applied. The predetermined contract value of demand, or billable demand (part of the demand rate), is sometimes referred to as the "ratchet clause." (When the demand meter registers a maximum demand, a ratchet prevents the register from moving in the direction of a lower value of maximum demand.)

Example 5.1. An example may serve to illustrate this concept. Assume a 100 horsepower (hp) motor operates 8 hours a day, 5 days a week, for a month (4 weeks or 20 days); its efficiency is 90%, and it is running at full load. The operating cost can be calculated:

$$kW_{used} = \frac{hp\ (0.746\ kW)}{Efficiency} \tag{5.1}$$

$$kW_{used} = 100 \times \frac{0.746}{0.90} = 83\ kW$$

To determine a monthly cost:

$$kWh - kW \times time$$

$$kWh = 83\ kW \times 8\ hr/day \times 20\ days = 13{,}280\ kWh$$

Assume rate schedule:

Consumer Service:	$10.00
Demand in excess of 10 kW:	$10.00/kWh
Energy:	
first 2,000 kWh	$0.15/kWh
next 3,000 kWh (add 100 kWh per kW > 10 kW)	$0.125/kWh
subsequent kWh	$0.10/kWh
Fuel Cost Adjustment	$0.01/kWh
Peak Contracted Demand	50 kW

Then:

Consumer Charge	$10.00
Demand Charge (83 kW – 10 kW) = 73 kW; @ $10/kW =	$730.00
Energy (2,000 kWh × $0.15/kWh) =	$300.00
Energy: (3,000 kWh × $0.125/kWh) =	$375.00

Energy/Demand: (100 kWh/kW × 73 kW) = 7,300 kWh;
(7,300 kWh × $0.125/kWh) = $912.50
Energy:
(13,280 kWh – 2,000 kWh – 3,000 kWh – 7,300 kWh)
= 980 kWh × $0.10/kWh $98.00
Fuel Adj.: 13,280 kWh × $0.01/kWh = $132.80
Total bill for electrical usage: $2,558.30

Note that the $730.00 direct charge for demand is almost 30% of the total bill. In addition, the second block of energy was extended by 7,300 kWh at a differential in rate ($0.125/kWh – $0.10/kWh) = $0.025/kWh or $182.50. Nearly 7% of the total bill may be considered a penalty for exceeding the 50 kW contract demand.

Power Factor Penalty

The power that is supplied to induction loads, through such means as induction motors, transformers, fluorescent lamps, and others, is not always fully effective. This is because the current and voltage do not act together (in phase with each other). The percentage of the effective power to the apparent power (voltage multiplied by amperes) is known as the power factor. The real or effective power is measured in watts or kilowatts (kW) while the apparent power is measured in volt-amperes, or kilovolt amperes (kVA). The component that causes the two quantities, volts and amperes, not to act in phase is known as the reactive power. It is also measured in volt-amperes or kilovolt amperes (kVA) and is known as the kilovar (kVAR). The reactive power produces the magnetic fields necessary for the operation of inductive devices. The vector sum (not the arithmetical sum) of the real and reactive power is the apparent power. (The vector sum is the net effect of two quantities acting 90° from each other, for purposes of measurement.)

Power factor, then, is the ratio of real power (kW) to apparent power (kVA):

$$\text{Power factor (\%)} = \left(\frac{\text{Real power (kW)}}{\text{Apparent power (kVA)}}\right) \times 100 \qquad (5.2)$$

The watt-hour meter measures only real power, but the generating, transmission, and distribution facilities must produce and provide for apparent power. The reactive power necessary for operation of inductive devices may be countered by reactive power produced by capacitors (the most common method). The installation of capacitors may make the real and apparent power coincide; this is known as power factor correction.

Power factor penalty rates attempt to have the consumer pay for the reactive power furnished or to invest in correcting equipment, a monthly charge vs. a one-time investment.

Other benefits result from power factor correction. Since less current flows for a given load, the voltage drop or loss will not only be less, causing loads to function better, but also reduce losses in the conductors. By the same token, supply equipment will not be loaded as much, providing capacity for growth.

Peak vs. Off-peak Pricing

As a means of reducing electric demands on their facilities, utilities offer special lower rates as incentives to consumers to operate their loads during off-peak periods. These generally fall into two major categories: time of day and time of year. Both depend on factors such as the prevailing seasonal weather, manufacturing demands, school sessions and vacations, others pertinent to particular areas.

Examples have been cited in chapter 3 in connection with lighting, heating, ventilation, and air conditioning loads. The causes and effects of such off-peak pricing are obvious. In many instances, prices may be raised during peak periods for similar reasons.

Demand Ratchet Clause

This is another attempt to control demand by specific rate provisions. Here the maximum demand is measured (usually by month) and applied throughout the succeeding months for a specified period (perhaps 12 months). It is charged as a minimum demand.

Cogeneration Power Cost Factor

When some of the power is supplied from nonutility-owned generating sources (cogeneration), the utility pays for it at a rate usually reflecting its least efficient generating units. (In some states, this is prescribed by law.) The difference in cost of energy thus supplied compared to that if supplied by the utility is shared by all the utility consumers and is included in the consumer bill. It is usually very small and is added to the metered kWh of the consumer; such a factor is in the nature of a few ten-thousandths of a cent per metered kWh.

Taxes

These usually include some form of sales taxes and income taxes. There may also be other taxes pertinent to the political jurisdiction in which the consumer lives or operates.

6

Electric Load Management and Deregulation

Electric load management under regulated utility operations and under deregulated marketers, resellers, and generation companies is basically the same technology. The common objective is to minimize rates for the ultimate consumer by increasing the electric system efficiency. The application and control of electric load management differs in several aspects under regulated and deregulated operations. These differences are explored in this chapter.

Basic Electric Load Management

Chapter 2 explained the components and relationships of load management as practiced by a utility supplying and delivering energy to its consumers. This includes:

- Researching consumer load characteristics
- Determining the consumer's contribution to system peak loads and annual energy requirements
- Selecting and implementing designs to establish load management

- Forecasting the effects of load management on system peaks and energy
- Using the forecasts to evaluate construction and operating needs
- Determining revenue requirements that are reflected in rates and financial planning of capital and operating costs
- Following up estimated results by establishing the actual results of load management strategies

Deregulation

Deregulation ostensibly encourages competition between utilities and other suppliers of electric energy for the consumers' benefit. Consumers are given the choice of suppliers for electric energy and associated services. Competition essentially forces each utility to improve efficiency so that prices and rate schedules may compete with other nonutility suppliers in the marketplace.

The drive for improved efficiency is the common objective of both regulated and deregulated electric load management. However, considering the consumer's reliance on electric energy, prudence dictates that the last ounce of fat in an electric system should not be squeezed out in the name of increased efficiency. An adequate margin of safety should always remain to prevent the occasional surge or overload from turning into a major service interruption.

Let's consider an area blackout (despite all the excellent post-mortem engineering that follows such incidents, they will continue to occur). When load diversity is lost, failure of some element during startup procedures may cascade, presenting operators with serious damage and service problems. Transmission systems are especially sensitive to such situations.

Electric systems have been historically divided into three functional parts—generation, transmission (the bulk power system), and distribution. Under deregulation, each of these parts may now be owned and operated by separate companies, with the bulk power system and generation being coordinated by an independent system operator (ISO). In some instances, a holding corporation may own these separate companies—often the parent of the original utility. Each functional part has its own particular challenges as it seeks to improve its competitive position.

During transformation from regulated to deregulated operations, some regulation must be maintained to control the effect of rates on consumers. Utilities will need time to supplant existing accounting practices peculiar to regulated operations with more general business accounting used by competitive industries.

Regulated utilities are currently mandated to serve all consumers in a nondiscriminatory manner within well-defined services areas. They operate under extensive regulation with respect to rates, finances, operations, environmental compliance, and service reliability standards. Regulated utilities make investment and operating decisions based on their assessment of current and projected consumer needs. Regulation is largely focused on rate schedules that allow for the recovery of prudently incurred costs plus a reasonable return on investment capital. The rate of return on capital, while limited by regulatory action, is not guaranteed to the investors.

Deregulation thus far has been left to the individual states, most of which use a phased—in process to introduce competition into the electric utilities within their borders. In addition to rate matters during the transition period, the utilities may reorganize the existing corporation. They may sell or reassign assets, retire or remove plants from service, and accelerate depreciation or amortization of assets with the approval of the states' regulatory authorities. Under deregulation, utilities are required to deliver electric service (via their transmission and distribution facilities) to those consumers who select a supplier other than the utility to whom they are connected. These consumers can now (or will soon be able to) receive their electric energy from any supplier, using the "home" utility's system along with any other intervening contiguous utility to the point of supply.

In theory, a consumer situated among other consumers on a utility system can apply for and receive less-expensive energy from a supplier located hundreds of miles away. In practice, however, charges added to the "less-expensive" energy for moving ("wheeling") this energy through the facilities of the connecting system and other intervening systems may make the choice of a remote supplier uneconomical. Add-on costs would include meter reading, service maintenance and reliability, and other services. For practical purposes, under deregulation the purchase and wheeling of "competitive" energy may be confined to large commercial and industrial customers or municipalities. (This was a common practice before deregulation and may remain so.)

Despite the lead taken by the states in these matters, the Federal Energy Regulatory Commission (FERC) is mandated to oversee the deregulation process, at least through the transition period. FERC orders 888 and

889, promoting competition between suppliers, require utilities to provide open access to their facilities on a nondiscriminatory basis. These orders mandate review of rates charged for such use. These rates help to cover certain "stranded" costs that are incurred by a utility for historic, prudent investments that are no longer useful under the new process. FERC may also mandate creation of ISOs, electronic control and information systems that share data on transmission reliability and capability and energy source agreements.

Successful implementation of electric industry deregulation rests largely on the capabilities of transmission systems, as they constitute the principal "highway" for the delivery of electric energy. In their efforts to be competitive, utilities may be forced to eliminate less-efficient generating units and transmission lines. They may also eliminate ancillary items necessary for service reliability. Such items might include redundant relay protection systems, line clearance maintenance, and research and development not necessary to direct operation of the system. Despite the sale of GPU's Three Mile Island plant to AmeriGen in 1998, as expensive nuclear generating plants are phased out, their large-capacity transmission outlets may no longer be needed. Lines, structures, and property could be sold or removed.

For economic, environmental, and public relations considerations, many transmission lines are located in remote areas, making them a very vulnerable link in the electric energy supply chain. In addition to exposure to the elements of nature and the vagaries of humans (vandalism, accidents), they are particularly susceptible to acts of sabotage. Despite alternate supply lines to a community, city, or large industrial complex, saboteurs with knowledge of an electric system can effectively deprive these consumers of electric service for extended periods of time. With local generation backup unavailable or curtailed to reduce costs for competitive reasons, the results of no electric service can have serious consequences (as in the 1965 Northeast blackout). Under wartime conditions, the consequences—especially to defense industries and activities—could be disastrous.

The fact that rates paid by commercial and industrial consumers "carry" residential consumers is well known. Under deregulation, the sizable revenues of commercial and industrial consumers could be diverted to remote companies because of better rate offers. If this occurs to any significant degree, the rates charged the remaining consumers (mostly residential) will be increased to make up for the lost revenue. This nullifies or offsets the principal reason for deregulation—to lower the cost of electric energy for all consumers.

Another possible scenario under deregulation is that lower energy costs cause the consumer to lose the inclination to participate in electric load management programs by the connecting electric utility. Yet, with declining generating plant capacity additions and fewer new bulk transmission lines, the continued development and use of load management remains a critical necessity. Another consideration is that the consumer will not realize immediate savings, since stranded costs are recoverable by the utility for a number of years.

Deregulation and the Telephone Industry

Comparisons with the deregulation of the telephone industry may be of interest. The action was supposed to improve service and lower consumer costs. Citing personal experiences may not equate with conclusions derived from a more scientific analysis, but they perhaps can serve as indicators of where such analysis should be made.

The regulated telephone companies issued a single monthly bill that was payable in person at the local telephone office. It was the single transaction between the supplier and the consumer. There are now at least two such bills—one from the local service company and one from the long-distance carrier. Instead of a single minimum charge, there are now multiple minimum charges. In addition, there is usually a fee charged by the local company for the "privilege" of having a connection made with the long-distance supplier. (Is this a charge soon to come for electric service?)

In most cases, bills can no longer be paid in person at the local telephone office but must be mailed to a distant collection center—at an additional cost. A local office, where consumers could register complaints or seek advice regarding their requirements for telephone service, has been replaced by central locations. Callers are faced with the button-pressing process to find specific services. Often, a caller endures a prerecorded message or is put on hold until a "service representative" is available. Woe to callers who incorrectly press the wrong buttons and have to start all over again! In the past, a request for directory assistance was a pleasant chat with another human being; the request is now usually computerized and results in an additional charge.

Has all this efficiency earned the consumer better service and reduced costs? Using the coin telephone as a rough yardstick—no! Before regulation, a local call cost 10¢. Under deregulation, a similar call is at least 35¢, an increase of 250%—greater by far than the cost of living increase during the comparable period. If the pay phone is not owned by the local company, the calls may have additional charges not immediately evident to the caller.

After two decades of experience in a deregulated environment, a number of telephone companies have sought mergers. Companies are combining to expand their areas of operation in order to reap benefits of scale and to better position themselves to meet competition. Four multibillion-dollar mergers were announced in the telecommunications industry in approximately the first half of 1998. If this trend continues, there will once again be a few giant communications companies serving the country—this time unregulated.

Some of these negative results and features have already been adopted by electric utilities. The increasingly competitive environment has resulted in some of them reorganizing and entering into mergers to achieve a better financial position in their market areas as they face an uncertain future.

Advances in the fields of electronic communications and automation have materially affected the daily activities of almost everyone. However, at a time when the demand for greater reliability and quality of electric service is more critical than ever, the potential for degrading service reliability by the deregulatory process is more than likely. This looming crisis should be the object of serious thought and action, yet service reliability is rarely, if ever, mentioned in the current deregulation process. Those concerned with establishing the deregulated electric power market should proceed with prudence and a degree of caution.

The fruits of electric load management are greater efficiency by electric utilities, whether regulated or unregulated. It's an advantage not easily disregarded.

7

The Integrated System

The planning, design, and operation of an electric utility system in past practice was a fairly straightforward process. It began with forecasting future load and planning the most economic means of meeting those demands by scheduling generation capacity additions and/or by increasing interconnection capacity for outside resources. This was done within a utility territory or on a regional basis through member companies in a power pool region. Construction of projects to meet the plan selected was expected to be completed within a reasonable time frame and at budgeted costs.

Changes began with the advent of environmental restrictions and public antipathy towards generating plant and transmission lines. This was due in part to concerns about appearance and possible health effects from electrical fields. Changes continued with the advent of deregulation, transmission access, and further environmental regulation. In addition were independent electric generation entities, corporate structure changes, and the potential for "green power" (environmentally acceptable energy sources) as well as distributed generation. During this time period some inescapable facts emerged:

- Energy requirements continue to grow (see Fig. 1–1)
- New generation capacity additions are declining (see Table 1–1)
- New transmission additions are declining (see Table 1–2)

In addition to the previously mentioned facts, the effects of independent power producers have to be considered in the planning and operation of electric power systems. In 1982, nonutility generation capacity was only 3% of the total electric utility generation capacity. By 1992, this increased to 7%, and in 1992, the total U.S. net new generation capacity increase was 6,830 megawatts (MW). Of this, nonutility sources (industrial cogenerators, independent power producers, and resource recovery plants) comprised 5,136 MW or 75% of the net new generating capacity.

Electric utility systems, both investor owned and public agencies, do not have significant control over the generation additions—or retirements—by nonutility entities. They have little control over the size, location, timing, design for reliability, mode of operation, or availability of capacity. There is no guarantee or assurance as to how long a nonutility generator will be in service if economies change, fuel supplies dwindle, or a lack of market for energy produced occurs because of competition from other sources.

The electric system requires integration of resources within the geographic system it supplies and with other systems in its region (power pools) in order to provide a reliable, economic electric energy source to consumers. There is virtually no area of life today not dependent upon this energy; every year, a new use is found. The word "reliability" is rarely found in the treatment of unrestricted transmission access, competition, deregulation, and restructuring. Yet reliability is a vital necessity in the production and delivery of electric energy. The consumer expects his electric energy source to be available whenever it's needed and in the quantities required.

Who Is Minding the Store?

- Investor-owned utilities and public power agencies are not planning and constructing new generation and transmission capacity to the same extent as in the past. How will new capacity be added? Who will take responsibility for it?

- With a narrowing margin between load and capability, what is and will be the effect on reliability? Who will determine generation capacity reliability under the increasing presence of nonutility generation, which is not subject to system or power pool requirements for reliability?

- What will be the effect on reliability of the sharply increasing number of power transactions handled hourly and daily by central system operators and power pool operators?

- What will take the place of nuclear station capacity being retired or closed prematurely by changes in economics or regulatory requirements? Conventionally fueled plants are to a lesser degree also affected by these changes, particularly air quality regulations.

- Who will decide the priority of allocating available power? How will it be allocated, both intrastate and interstate? Will it be decided by economics alone, or by need? For instance, an element of an interconnection outlet might suffer damage due to a fault and is taken out of service. This reduces the amount of energy available for consumers of remote suppliers on the connecting utility system. Who determines which consumers get reduced in energy supply, such as municipal customers, large industries, or individual consumers of supplier A, B, or C?

- Under deregulation, if a consumer is supplied energy through an electric system owned and operated by another entity, where does the responsibility lie for metering and billing the consumer? In other competitive industries that meter delivery to the consumer (heating fuel, gasoline, mail, etc.), the supplier meters the delivery.

- Should the electric system delivering energy to the consumer from an energy service be responsible for theft of service detection and prosecution?

- Will a remote energy supplier to a consumer be responsible for connecting or disconnecting service? Will the supplier be responsible for reading the meter at the time of change from the connecting utility supplier to a remote supplier?

- Will the delivering system have access to consumer's energy and demand usage that is supplied energy by a remote source? Will duplicate accounting systems be necessary? Who will supply consumer usage data to data collecting agencies such as the Department of Energy (DOE) and the Edison Electric Institute (EEI)?

Some of the questions posed above may be answered by remote meter reading systems such as those developed by American Innovations, which also offers an independent service for database and billing. With such systems, a consumer may have the meter read at any time and the billing information known to the existing supplier without a field visit by a meter reader.

These and many other questions remain to be answered as the electric utility industry enters the next century.

The Role of Electric Load Management

Electric load management has been an effective tool in mitigating the disparity between increasing electric usage and the decreasing net additional capacity to supply loads. It has been and will continue to be effective when integrated into system planning of resources and system operation. Evaluation of electric load management will be increasingly important under the changing conditions encountered in the electric utility sector. (See Runnels and Whyte, "Evaluation of DSM," *EEE Proceedings*, Vol. 73, No. 10, 1985.) The different levels of complexity will have to be coordinated and integrated for individual, power pool, and regional systems.

Appendix A

The Electric Utility System

The typical electric utility system is comprised of generation, transmission, substation, and distribution components. They are operated together to supply electrical energy to consumers at any given time and in any given amount of energy required (see Fig. A–1).

Generation

Generation consists of the electrical energy producing or manufacturing facilities owned by the utility company and those facilities owned by others that are connected to the electric utility system. The latter may be in the form of refuse recovery plants, cogeneration, or separately owned generating plants. A utility may also own all or part of a generating plant located remotely from its own territory.

Because consumer use of electricity fluctuates, electric utility generating units are designed to handle three major types of load requirements the utility must meet—base load, intermediate load, and peak load.

A *base load generating unit* is used to satisfy all or part of the minimum or base load of the system. As a consequence, it produces electricity at an essentially constant rate and runs continuously. Base load units are the newest, largest, and most efficient of the three types of units.

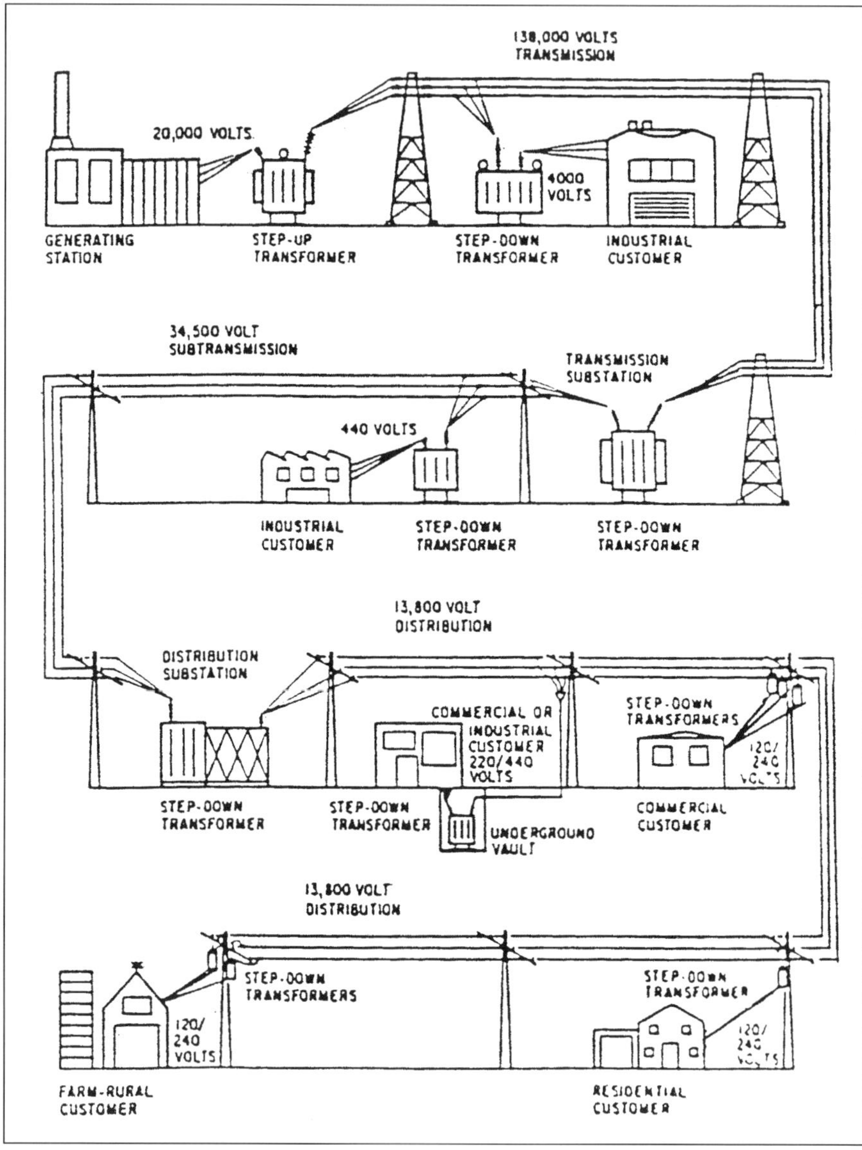

Fig. A–1 Electric supply from generator to customer

A *peak load generating unit*, which is normally the least efficient of the three unit types, is used to meet requirements during the periods of greatest or peak load on the system. Intermediate load generating units meet system requirements that are greater than base load but less than peak load.

Intermediate load units are used during the transition between base load and peak load requirements. Utilities also have reserve or standby generating units, which are available to the system, should an unexpected increase in load or an unexpected outage within the system occur. This is the reason that electric utilities must account for reserve or standby capability in a standard inventory of net capability, as well as account for generating units that are not available to the system, for such reasons as routine maintenance.

Generating facilities are usually designated as either *base load generation* or *peaking generation* based on their assignment. Base load generation units—generally the most economical energy-producing units—run to supply a major portion of the system load for most of the time. Peaking generation units are run to make up the difference between base load and the maximum loads either during daily load cycles or seasonally.

The Fundamentals of Power Generation

The majority of the electricity generated throughout the world is the product of burning nonrenewable fossil fuels—coal, oil, natural gas, or nuclear fuel—or from the use of renewable fuel resources, such as the force of water (known as hydroelectricity). Other sources of fuel are, in the main, derivatives of these fuels and processes (such as the burning of biomass, usually the waste products from the agriculture industry). Coal is by far the primary energy source used by electric utilities to produce electricity in the United States, with a share of around 55%.

However, production of electricity from any one of these energy sources is not significantly different from the others. For purposes of production of electricity, any one of them could be represented in the "generating station" in Figure A-1. For purposes of discussion, then, the type of power plant illustrated in Figure A-1 is that of a standard steam power plant that converts primary fuel (fossil, nuclear, etc.) into electricity.

The fuel heats water in the boiler, turning the water into steam; the steam turns the turbine and the generator. An electromagnet located inside the generator's stationary coil of wire is surrounded by magnetic lines of force that travel from one end of the magnet to the other. An electric current is established in the wire when the magnetic field produced at the ends of the magnet moves across the turns of wire in the stationary coil. When the magnet turns, the lines of force it creates cuts through the wire and thus induces electric current into the wire. This process generates the electricity that then goes into the overall electricity network.

Improving Generation Efficiency

The task at hand in this book, of course, has been improving load management. Germane to any discussion of load management is the problem inherent in the entire generation process: It's not very efficient. More than half of the heat (steam) in the fuel escapes through a smokestack as exhaust heat or wasted steam.

To try and limit this problem, two procedures have been developed to get as much use as possible from the fuel: *combined-cycle generation* and *cogeneration*.

Combined-Cycle. A *combined-cycle plant* combines a gas turbine and a steam turbine. As the gas turbine produces electricity, it passes waste heat to the steam turbine, which also produces electricity. This process reduces the amount of fuel needed to heat the water to reach the temperature needed to produce steam. Generally, a steam turbine may only convert 35% of the heat content of fuel to electricity (35% efficiency); a combined-cycle plant can reach 50% efficiency.

Cogeneration. This technology is found mainly among nonutility generators. It is the combined production of electric power and a second form of energy (such as heat or steam) through the use of a single energy source. The process can begin either with heat or steam production or with electricity production. Unused energy from the first process is used as input to the second process. The primary energy source is usually a fossil fuel, although renewables can also be used.

Various technologies may be employed to produce both electric power and another form of useful energy using the cogeneration process; however, the cogenerating technologies are classified either as *topping-cycle systems* or *bottoming-cycle systems*. How it is classified depends upon whether electrical or thermal energy is produced first.

In *topping-cycle facilities*, energy input to the system is generally first transformed into electricity by using high-temperature, high-pressure steam from a boiler to drive a turbine. Waste heat—the lower-pressure steam exhausting from the turbine—is used as a source of processed heat. Topping-cycle systems are the most common and are used in commercial and industrial applications.

In a *bottoming-cycle system*, high-temperature thermal energy—steam—is produced first for commercial or industrial applications. Heat extracted from the hot exhaust stream is then carried off to drive a turbine. Industrial processes that require very high temperature heat—thus making it economical to recover the waste heat—generally use bottoming-cycle systems.

Fossil-fueled steam turbine systems are used in most industrial cogenerating processes, while gas-turbine systems are used in most utility and nonutility processes.

Distributed Generation. Yet another method of increasing overall system efficiency is *distributed generation*—a development that is bringing substantial and dramatic changes to the way the generation industry has been operated in modern times.

Distributed generation means that the production of electricity is no longer solely dependent upon large, remote utility facilities. Instead, smaller customer-owned generating units provide back-up or peaking power (or even total, self-contained power supplies, in limited cases). Distributed generation can facilitate load management by improving the reliability of the electric power supply, managing system-wide peaks in demand, and deferring the cost of new power plants and lines. Because the time between planning and operating a major power plant can involve between five and seven years (on average)—with many more years required for payoff of the investment—distributed generation systems can be up and running within one year and paid off in between three and five years.

EPRI classifies distributed generation facilities as those with a capacity of between 5 and 100 MW. Generating technologies include diesel and gasoline-fired internal combustion; fuel cell systems; microturbines, and photovoltaics. In practice, generator sets are installed at customer facilities and operated during what are peak demand periods for the customer's utility supplier. Customers avoid peak power consumption and so minimize demand charges, reaping immediate cost-savings and avoid load curtailment. Utilities can cost-effectively meet demand and more efficiently utilize their generation and transmission systems.

Diesel engine systems, while otherwise limited in cogeneration applications (since they provide less useable processed heat per unit of electric power output), are attractive to small cogeneration or distributed generation applications that need instantaneous supply of electricity where the electric power requirement is generally greater than the heat requirement. In a diesel system, the engine is cooled with water. The heated water is then used for processed steam, heat, or hot water applications. Exhaust gases can be used in a similar manner.

To assess the best potential use of distributed generation a facility manager should work together with his utility supplier and assess the following:

- Load usage patterns
- Fuels costs (and which fuels are primarily used)
- Transmission and distribution system reliability
- Distributed generation technology available
- Up-front capital costs to build and start-up a system
- Annual maintenance costs of the equipment
- Avoided transmission and distribution costs
- What the savings would be over purchased electricity costs

Proviso: The authors wish to note that it has not been demonstrated that distributed generation is a method of increasing overall system efficiency. For instance, the diversity lost because a distributed generator must be sized to meet an individual peak load and not the lower diversified peak as seen by central generation has not been quantified as to cost in any resource that we have seen thus far. A typical residential house load of 3 kW that its distributed generator must supply is seen only as 1.8 kW at a large central generator using a coincidence factor of 60%. This difference on a *system* basis has not been mentioned by proponents of distributed generation.

In addition to cost difference, what is the effect on emissions for the additional distributed generation capacity *not* currently subject to stringent regulation and inspection, as are central units? What about cumulative noise effect in residential neighborhoods compared to central units at a remote location, again subject to stringent regulation?

As to reliability, unless a customer has an alternate electric supply such as another generator, or relies on a backup supply from the utility, he is subject to maintenance outages, equipment breakdowns, and disruption of fuel supply (through truck strikes, shortages, etc.) Where will all the service personnel come from? What are the costs involved? If a utility backup is required, the system capacity must be sized to meet these loads as well as the normal utility loads.

Transmission

Once electric power is generated, it must be transmitted to consumers. The transmission part of the electric system moves the power from generating plants and interconnections with other utilities to load centers throughout the utility territory. Through a network of components, it is the transmission and distribution (T&D) system's responsibility to deliver the electricity in the quantity that is needed to meet particular customer demands.

Transmission lines operate at voltages up to 750,000 volts (V) AC and may be overhead on structures or underground, and in some special cases, under water.

At load centers in the utility territory, the transmission voltage is reduced (stepped down) to a primary distribution voltage or in some cases to a subtransmission voltage. This allows power to be brought to consumers in the local area by the primary distribution system.

Power Transmission Elements

The components of a typical T&D system include a switchyard, transmission lines, a substation, and distribution lines. On the transmission side of a T&D system, transmission lines and a utility's switchyard are the major components.

A utility's *switchyard* is typically located adjacent to the power plant. It receives electricity the power plant generates, directing it to the *transmission lines* and, by means of those lines, to the *power grid*. A power grid is a network of major electrical conductors that link together individual utilities. Such a power grid is essential for those times when customers' demand for power exceeds what an individual electric utility company can produce. The grid ensures a continuous flow of power to consumers at all times.

Transmission lines carry power from the switchyard to *substations* via *conductors*—materials offering minimal resistance to the flow of electricity. Transmission lines also have connections between the substations to provide more than one way for the power to get to individual substations. When a substation receives electricity from the transmission lines, it transforms it (as necessary) to supply a given distribution area.

Customers ultimately receive electricity from distribution lines—the direct link between them and the rest of the electric system.

Primary Distribution

From the substation, power is delivered to the local area over primary voltage distribution lines operating at voltages from 4,000 V to as many as 45,000 V. The most frequently used primary voltage range is 13,000 V. As in the case of transmission, distribution lines may be overhead or underground, but it is much more frequently used underground than in transmission systems. This is particularly true in urban areas and high load density locations, such as shopping centers or industrial parks.

Distribution Transformers

A distribution transformer is used to reduce the primary voltage to a utilization voltage that consumers can use within their premises. Residential consumers are usually served at 240/120 V. Appliances such as large air conditioners, electric ranges, and pumps utilize 240 V, while lighting, TVs, and computers use 120 V. Commercial and industrial consumers use higher utilization voltages (up to 480 V).

Distribution Systems

Distribution lines serve two basic functions. They carry power from a substation to consumer areas, and deliver power to consumers.

Distribution systems are said to be physically installed either *overhead* or *underground*. In an overhead distribution system, power is delivered through conductors that are strung from pole to pole. Underground systems deliver power through cables that are buried or are run through conduit or ducts. Overhead and underground systems typically use many of the same components. Overhead systems are generally more common (largely because they are less expensive to construct and maintain).

Most often, wood poles are used in these systems to support insulators and conductors. Insulators perform the same function on distribution lines as they do on transmission lines—isolating conductors to prevent them from making contact with the pole or other conductors.

Most insulators are classified as *pin-type*. Mushroom-shaped pin-type insulators are usually made of porcelain and are attached to a threaded pin, which is bolted to a crossarm on the pole. Insulators are generally used to support the primary feeder conductors, which link the distribution system to the substation.

A *distribution transformer*, mounted on a pole, reduces voltage from distribution levels to levels that are suitable for consumers. The *primary bushings*, which are located on the top of the transformer, are connected to the primary main at the top of the pole. The distribution transformer reduces the typical 13.8 kilovolt primary voltage to 120/240 volts, which is a voltage commonly used by residential consumers.

The lower-voltage electricity passes through the transformer's *secondary bushings* to the *secondary conductors*. These secondary conductors are usually strung to poles on either side of a transformer to supply more customers in areas where homes are close together. From the secondary conductors, *service drops* are run directly to residences from the secondary main conductors, delivering electricity directly to homes and businesses. In sparsely populated rural areas, service drops deliver power to customers directly from the distribution transformer. Large commercial and industrial customers also may have service directly from a distribution transformer.

It is the secondary distribution lines that ultimately deliver the electricity to consumers in accordance with the proper voltage.

Fused switches interrupt the circuit and stop current flow should an electrical problem arise in the transformer or secondary circuit. Switches can also be used to open or close circuits when equipment has to be taken out of service for maintenance purposes. They are generally found on the primary system.

As can be seen, generation, transmission, and distribution are inextricably bound to one another and to any discussion of load management on an electric utility system.

Appendix B

Economic Studies

Economic studies are the means of evaluating the economic consequences of a particular proposal or a number of alternate proposals for meeting a problem. They may range from the very simple to the very complex; in some instances, they may appear to be the application of good common sense. Several methods for economic evaluation are:

1. Simple payback period
2. Breakeven period
3. Benefit-to-cost ratio
4. Net annual savings
5. Return on investment
6. Life-cycle cost analysis

The first four methods listed are essentially the same, but the results are expressed in a different manner.

Simple Payback Period

The payback period may be expressed as a ratio:

$$\text{Simple payback period} = \frac{\text{Total first cost of project}}{\text{First year savings}}$$

Example B.1. If the total cost of an economizer installed on an air conditioning system to contribute to a hot water supply system is \$10,000, and reduction in heating cost for the first year of operation is \$2,000, then the simple payback period is:

$$\text{Simple payback period} = \frac{\$10{,}000}{\$2{,}000/\text{yr}} = 5 \text{ years}$$

Breakeven Period

The breakeven period is the time at which the savings realized will equal the expenditure necessary to achieve the savings. For the example shown, the obvious answer is five years; that is, the annual savings must be realized for at least five years to justify the expenditure.

Benefit-to-cost Ratio

This is the inverse version of the payback period, without reference to time:

$$\text{Benefit-to-cost ratio} = \frac{\text{Benefit (savings at one time or other)}}{\text{Total first cost of the project}}$$

For the example, the ratio for the first year would be \$2,000/\$10,000, or 1 in 5.

These methods do not take into account other considerations such as maintenance and operation, taxes, utility rate, interest, or the time value of money. However, their simplicity makes them useful not only for relatively small projects, but as rough indicators whether further, more complex methods of study are warranted.

Net Annual Savings

Here the estimated incoming cash flow is reduced by the estimated outgoing cash flow:

Net annual savings = incoming cash flow – outgoing cash flow
Incoming cash flow = estimated annual savings + annual tax credit*
(*For a one-time credit, the equivalent annual value over the estimated life of the project may be used.)

Outgoing cash flow = annual debt charge** + annual operating and maintenance costs

[**Debt charge is the first cost of the project (including capital equipment + labor + material) multiplied by a uniform capital recovery factor (see Table B–1)]

Uniform capital recovery factor (present worth to annuity):

$$\text{Uniform capital recovery factor} = \frac{i(1+i)^{n}}{(1+i)^{n-1}}$$

where

i is the interest rate
n is the number of years (estimated life of project equipment)

It should be noted that:

- Estimated annual savings depend on weather, energy and demand, rate schedules, and other potential variables
- Annual tax credits depend generally on legislation
- Annual operating and maintenance costs depend on inflation rates

The individual items in the preceding equations are usually taken as average values over the estimated life of the project. If more accuracy is desired, the net annual savings may be estimated for each of the years in the established life of the project.

Return on Investment

This method is essentially an expansion of the net annual savings method and serves to indicate not only the quantity of the savings but their quality as well. It is usually expressed as a percentage. The first cost of the project includes capital equipment, labor, and material × 100.

$$\text{Return on investment (\%)} = \frac{\text{Net annual savings}}{\text{First cost} \times 100}$$

Table B-1 Uniform Capital Recovery Factor

Yr:	2%	4%	6%	8%	10%	12%	15%	20%	25%	30%
1	1.020	1.040	1.060	1.080	1.100	1.120	1.150	1.200	1.250	1.300
2	.5150	.5302	.5454	.5608	.5762	.5917	.6151	.6545	.6944	.7348
3	.3468	.3603	.3741	.3880	.4021	~4163	.4380	.4747	.5123	.5506
4	.2626	.2755	.2886	.3019	.3155	.3292	.3503	.3863	.4234	.4616
5	.2122	.2246	.2374	.2505	.2638	.2774	.2983	.3344	.3719	.4106
6	.1785	.1908	.2034	.2163	.2296	.2432	.2642	.3007	.3388	.3784
7	.1545	.1666	.17g1	.1921	.2054	.2191	.2404	.2774	.3163	.3569
8	.1365	.1485	.1610	.1740	.1874	.2013	.2229	.2606	.300~	.3419
9	.1225	.1345	.1470	.1601	.1736	.1877	.2096	.2481	.2888	.3312
10	.1113	.1233	.1359	.1490	.1627	.1770	.1993	.2385	.2801	.3235
11	.1022	.1141	.1268	.1401	.1540	.1684	.1911	.2311	.2735	.3177
1?	.0946	.1066	.1193	.1327	.1468	.1614	.1845	.2253	.2685	.3135
13	.0881	.1001	.1130	.1265	.1408	.1557	.1791	.2206	.2645	.3102
14	.0826	.0947	.1076	.1213	.1357	.1509	.1747	.2169	.2615	.3078
15	.0778	.0899	.1030	.1168	.1315	.1469	.1710	.2139	.2591	.3060
16	.0737	.0858	.0990	.1130	.1278	.1434	.1679	.2114	.2572	.3046
17	.0700	.0822	.0954	.1096	.1247	.1405	.1654	.2094	.2558	.3035
18	.0667	.0790	.0924	.1067	.1219	.1379	.1632	.2078	.2546	.3027
19	.0638	.0761	.0896	.1041	.1195	.1358	.1613	.2065	.2537	.3021
20	.0611	.0736	.0872	.1019	.1175	.1339	.1598	.2054	.2529	.3016
21	.0588	.0713	.0850	.0998	.1156	.1322	.1584	.2044	.2523	.3012
22	.0566	.0692	.0830	.0980	.1140	.1308	.1573	.2037	.2519	.3009
23	.0547	.0673	.0813	.0964	.1126	.1296	.1563	.2031	.2515	.3007
24	.0529	.0656	.0797	.0950	.1113	.1285	.1554	.2025	.2512	.3006
25	.0512	.0640	.0782	.0937	.1102	.1275	.1547	.2021	.2510	.3004
30	.0.446	.0578	.0726	.0888	.1061	.1241	.1523	.2008	.2503	.3001
35	.0400	.0536	.0690	.0858	.1037	.1223	.1511	.2003	.2501	
40	.0366	.0505	.0664	.0839	.1023	.1213	.1506	.2001	.2500	
45	.0339	.0483	.0647	.0826	.1014	.1207	.1503	.2001		
50	.0318	.0466	.0634	.0817	.1009	.1204	.1501	.2000		

Table B-2 Future Worth To Present Worth Factor

Yr:	2%	4%	6%	8%	10%	12%	15%	20%	25%	30%
1	.9804	.9615	.9434	.9259	.9091	.8929	.8696	.8333	.8000	.7692
2	.9612	.9246'	.8900	.8573	.8264	.7972	.7561	.6944	∑6400	.5917
3	.9423'	.8890	.8396	.7~38	.7513	.7118	.6575	.5787	.5120	.4552
4	.9238	.8548	.7921	.7350	.6830	.6355	.5718	.4823	.4096	.3501
S	.9057	.8219	.7473	.6806	.6209	.5674	.4972	.4019	.3277	.2693
6	.8880	.7903	.7050	.6302	.5645	.5066	.4323	.3349	.2621	.2072
7	.8706	∑759g	.6651	.5835	.5132	.4523	.3759	.2791	.2092	.1594
8	,8535	.7307	.6274.	.5403	.4665	.4039	.3269	.2326	.1678	.1226
9	.8368	.7026	.5919	.5002	.4241	.3606	.2843	.1938	.1342	.0943
10	.8203	.6756	.5584	.4632	.3855	.3220	.2472	.1615	.1074	.0725
11	.8043	.6496	.5268	.4289	.3505	.2875	.2149	.1346	.0859	.0558
12	.7885	.6246	.4970	.3971	.3186	.2567	.1869	.1122	.0687	.0429
13	.7730	.6006	.4688	.3677	.2897	.2292	.1625	.0935	.0550	.0330
14	.7579	.5775	.4423	.3405	.2633.	.2046	.1413	.0779	.0440	.0254
15	.7430	.5553	.4173	.3152	.2394	.1827	.1229	.0649	.0352	.0195
16	.7284	.5339	.3936	.2919	.2176	.1631	.1069	.0541	.0281	.0150
17	.7142	.5134	;3714	.2703	.1978	.1456	.0929	.0451	.0225	.0116
18	.7002	.4936	.3503	.2502	.1799	.1300	.0808	.0376	.0180	.0089
19	.6864	.4746	.3305	.2317	.1635	.1161	.0703	.0313	.0144	.0068
20	.6730	.4564	.3118	.2145	.1486	.1037	.0611	.0261'	.0115	.0053
21	.6698	.4388	.2942	.1987	.1351	.0926	.0531	.0217	.0092	.0040
22	.6468	.4220	.2775	.1839	.1228	.0826	.0462	.0181	.0074	.0031
23	.6342	.4057	.2618	.1703	.1117	.0738	.0402	.0151	.0059	.0024
24'	.6217	.3901	.2470	.1577	.1015	.0659	.0349	.0126	.0047	.0018
25	.6095	.3751	.2330	.1460	.0923	.0588	.0304	.0105	.0038	.0014
30	.5521	.3083	.1741	.0994	.0573	.0334	.0151	.0042	.0012	
35	.5000	.2534	.1301	.0676	.0356	.0189	.0075	.0017	.0004	
40	.4529	.2083	.0972	.0460	.0221	.0107	.0037	.0007	.0001	
4S	.4102	.1712	.0727	.0313	.0137	.0061	.0019	.0003		
50	.3715	.1407	.0543	.0213	.0085	.0035	.0009	.0001		

Example B.2. For example, consider the installation of a programmable time clock costing $100 that will result in estimated net annual savings of $400.

$$\text{Return on investment} = \left(\frac{\$400}{\$100}\right) \times 100 = 400\%$$

The calculations indicate that this apparently is an important project to consider further.

However, let's compare this to an economizer installed at an air conditioning unit costing $20,000, which will result in an estimated net annual savings of $4,000:

$$\text{Return on investment (ROI)} = \left(\frac{\$4{,}000}{\$20{,}000}\right) \times 100 = 20\%$$

Obviously, the $4,000 savings are more important than the $400. However, if many time clocks can be installed at a large number of units, the savings will be as significant as those from the single economizer.

Payback periods can also be established from this method.

$$\text{Payback period (years)} = \frac{100\%}{\text{ROI (\%)}}$$

For the previously given data:

$$\text{Payback of clocks} = \frac{100}{400} = \frac{1}{4} \text{ of a year} = 3 \text{ months}$$

$$\text{Payback of economizer} = \frac{100}{20} = 5 \text{ years}$$

It should be noted that in both methods, the net annual savings and return on investments are important financial criteria.

Life-cycle Analysis

This method may be applied to any type of capital investment that involves cost factors; often higher first costs may have to be justified by lower life-cycle costs.

The net annual savings and return on investment methods may be viewed as simple forms of life-cycle cost analysis because they take into consideration time value of money and future cost elements. However, the life-cycle cost analysis method translates all cost elements from the future to the present time and combines them with the first cost. Hence, it is sometimes referred to as the present worth method, although the reverse may sometimes be desired, i.e., future worth of present expenditures (see Table B–2).

Time Value of Money

Earning power

Money has earning power. A dollar today is worth more than a dollar a year from now because of this earning power available through investment. The precise value of today's dollar in the future will depend upon the rate of interest earned on the invested dollar. Thus, $1.00 today, invested at a 7% interest rate, will be worth $1.07 one year in the future. Conversely, $1.07 available a year from now has a present worth of $1.00. By using the concept that money has an increasing value over a period of time, any expenditure in the future may be expressed in its equivalent "present worth" today. This principle is used to convert expenditures made at varying times to an equivalent value at any one given instant.

Conversions

Such conversions may be made by converting values to:

1. Present worth—the value today
2. Future worth—the value at any specified time in the future
3. Annuity—a uniform series of payments over a period of time

The result of spending capital money is a series of annual charges extending over the service life of the property in which the capital is invested. Some of these annual charges will be uniform every year and may be considered an annuity. Other annual charges will vary from year to year resulting in a nonuniform series; these can be converted to a uniform series. Conversion factors at 7% interest for all these manipulations are provided in Table B–3.

Table B-3 Compound Interest

	Lump sum		*Uniform annual series*			
n	*Present worth to future worth,* $(1+i)^n$	*Future worth to present worth,* $\frac{1}{(1+i)^n}$	*Annuity to future worth* $\frac{(1+i)^n-1}{i}$	*Future worth to annuity* $\frac{i}{(1+i)^n-1}$	*Annuity to present worth* $\frac{(1+i)^n-1}{i(1+i)^n}$	*Present worth to annuity* $\frac{i(1+i)^n}{(1+i)^n-1}$
1	1.070	0.9346	1.000	1.00000	0.935	1.07000
2	1.145	0.8734	2.070	0.48309	1.808	0.55309
3	1.225	0.8163	3.215	0.31105	2.624	0.38105
4	1.311	0.7629	4.440	0.22523	3.387	0.29523
5	1.403	0.7130	5.751	0.17389	4.100	0.24389
6	1.501	0.6663	7.153	0.13980	4.767	0.20980
7	1.606	0.6227	8.654	0.11555	5.389	0.18555
8	1.718	0.5820	10.260	0.09747	5.971	0.16747
9	1.838	0.5439	11.978	0.08349	6.515	0.15349
10	1.967	0.5083	13.816	0.07238	7.024	0.14238
11	2.105	0.4751	15.784	0.06336	7.499	0.13336
12	2.252	0.4440	17.888	0.05590	7.943	0.12590
13	2.410	0.4150	20.141	0.04965	8.358	0.11965
14	2.579	0.3878	22.550	0.04434	8.745	0.11434
15	2.759	0.3624	25.129	0.03979	9.108	0.10979
16	2.952	0.3387	27.888	0.03586	9.447	0.10586
17	3.159	0.3166	30.840	0.03243	9.763	0.10243
18	3.380	0.2959	33.999	0.02941	10.059	0.09941
19	3.617	0.2765	37.379	0.02675	10.336	0.09675
20	3.870	0.2584	40.995	0.02439	10.594	0.09439
21	4.141	0.2415	44.865	0.02229	10.836	0.09229
22	4.430	0.2257	49.006	0.02041	11.061	0.09041
23	4.741	0.2109	53.436	0.01871	11.272	0.08871
24	5.072	0.1971	58.177	0.01719	11.469	0.08719
25	5.427	0.1842	63.249	0.01581	11.654	0.08581
26	5.807	0.1722	68.676	0.01456	11.826	0.08456
27	6.214	0.1609	74.484	0.01343	11.987	0.08343
28	6.649	0.1504	80.698	0.01239	12.137	0.08239
29	7.114	0.1406	87.347	0.01145	12.278	0.08145
30	7.612	0.1314	94.461	0.01059	12.409	0.08059

There are a number of different ways of developing the conversion factors. The convention used in Table B–3 is that annuity payments and future worth values are evaluated at the end of periods (years) and present worth values are evaluated at the beginning of periods. Developed in this way, Table B–3 is in its most directly usable form, since all payments are assumed to be made at the end of a calendar year (December 31) throughout this study.

Eight conversions cover all cases and are summarized and illustrated in the following examples and worth-time diagrams (courtesy Long Island Lighting Company).

Example B.3. The present worth is converted to future worth calculated using a single amount at any date to single amount at any subsequent date.

You have just won $5,000, tax free. How much money will you have at the end of 10 years, if you invest it at 7%, compounded annually?

In order to determine this amount, first refer to Table B–1. The $5,000 is a present worth, the value 10 years hence is a future worth. The future worth is obtained by multiplying the present worth by the conversion factor "present worth to future worth" for 10 years from Table B–3:

Future worth in 10 years = $5,000 × 1.967 = $9,835

Example B.4. Calculate present worth from future worth using a single amount at any date to single amount at any previous date.

You have estimated that 10 years from now the unpaid mortgage on your house will be $9,835. How much money do you have to invest today at 7% interest to just accumulate $9,835 in 10 years?

The future worth is $9,835; the present worth of that amount can be determined using Table B–3. From Table B–3, we learn that the conversion factor is 0.5083. Consequently:

Present worth = $9,835 × 0.5083 = $5,000

This is the reverse of Example B.3. The conversion factor for future worth to present worth is simply the reciprocal of the present worth to future worth factor. The worth-time diagram is the same as for Example B.3. (Fig. B-1a)

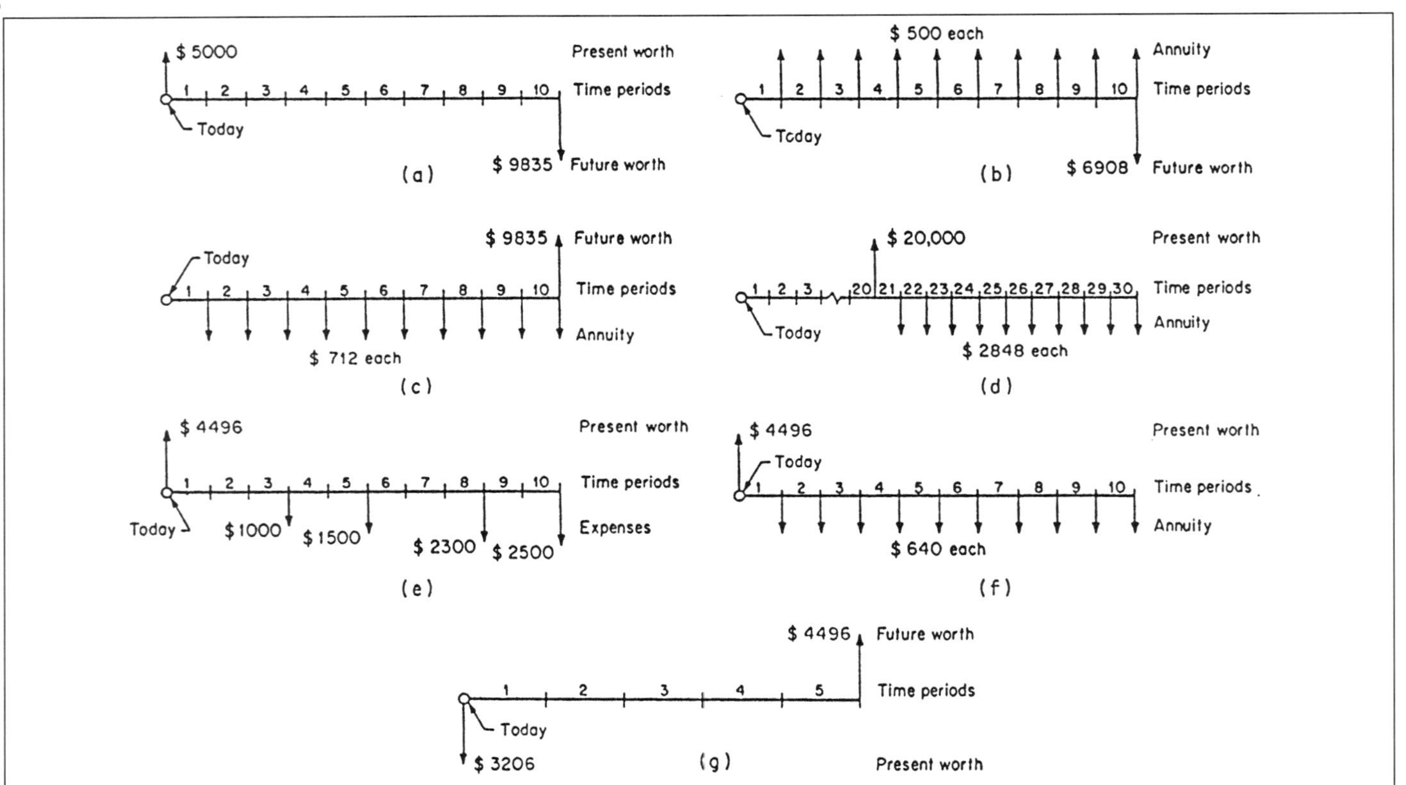

Fig. B-1 (a) Illustrating present worth to future worth. **(b)** Illustrating annuity to future worth. **(c)** Illustrating future worth to annnuity. **(d)** Illustrating present worth to annuity. **(e)** Illustrating non-uniform expense. **(f)** Illustrating uniform annual charge. **(g)** Illustrating present worth, years hence vs. today.

Example B.5. Calculate annuity to future worth (using an annuity over any period to single amount at end of period).

You plan to save $500 of your earnings each year for the next 10 years. How much money will you have at the end of the tenth year if you invest your savings at 7% per year?

The $500 each year is an annuity, since it is a uniform amount each year. You wish to know the future worth. From Table B–3, the annuity to future worth factor, 10 years, is 13.816:

$$\text{Future worth of the annuity} = \$500 \times 13.816 = \$6{,}908$$

Note from Figure B–2 that annuity payments are assumed to be made at the end of each time period. The conversion factor evaluates future worth at the same time that the last annuity payment is made. (Fig. B-1b)

Example B.6. Calculate future worth to annuity (using a single amount at any given date to annuity over any previous period ending at the given date).

If the unpaid mortgage on your house in 10 years will be $9,835, how much money do you have to invest annually at 7% interest to have just this amount on hand at the end of the tenth year? (Fig. B-1c)

First, refer to Table B–3. The $9,835 is a future worth; the uniform amount (annuity) to set aside annually is desired. From Table B–3, the future worth to annuity factor, 10 years, is 0.7238. Consequently:

$$\text{Annuity} = \$9{,}835 \times 0.07238 = \$712$$

Example B.7. Calculate present worth to annuity (using a single amount at any given date to annuity over any subsequent period starting at the given date).

You hold an endowment-type insurance policy that will pay you a lump sum of $20,000 when you reach age 65. If you invest this money at 7% interest, how much money can you withdraw from your account each year so that at the end of 10 years, there will be nothing left? (Fig. B-1d)

The $20,000 can be considered the present worth at the end of the 10th year. From Table B–3, the present worth to annuity factor, 10 years, is 0.14238:

Annuity (withdrawn for 10 years) = \$20,000 × 0.14238 = \$2,848

Note that direct use of the conversion factor assumes the first withdrawal takes place one period after the lump sum of \$20,000 is received.

Example B.8. Calculate annuity to present worth (using annuity over any period to single amount at start of the period).

You have estimated that for the first 10 years after you retire you will require an annual income of \$2,848. How much money must you have invested at 7% at age 65 to realize just this annual income?

The present worth of an annuity for 10 years is desired. From Table B–3, the annuity to present worth factor, 10 years, is 7.024:

Present worth = \$2,848 × 7.024 = \$20,000

The worth-time diagram is the same for Example B.7. (Fig. B-1d)

Example B.9. Convert present worth nonuniform expenses to equivalent uniform annual charge.

The maintenance expenses for the next 10 years on a piece of equipment are estimated as follows:

Year	Amount	Factor Fw to Pw	Percent Worth (Pw)
3	\$1,000	0.8163	\$816
5	\$1,500	0.7130	\$1,070
8	\$2,300	0.5820	\$1,339
10	\$2,500	0.5083	\$1,271

What is the present worth of these expenses? What is the uniform annual payment for 10 years equivalent to this nonuniform series? What does this mean?

In order to arrive at the solutions, we need to refer to Figure B–1e. The expense amounts are future worths in the year indicated. The present worth is desired. We can use future worth to present worth factors from Table B–3. Calculating for each year, we arrive at present worths for years 3, 5, 8, and 10 of \$816, \$1,070, \$1,339, and \$1,271, respectively, for a total worth of this nonuniform series expenses of \$4,496.

The equivalent uniform annual series is obtained by applying the present worth to annuity factor for 10 years to the present worth. From Table B–3, present worth to annuity factor, 10 years, is 0.14238. Consequently:

Equivalent uniform annual charge = \$4,496 × 0.14238 = \$640

This means that if you had \$4,496 and invested it at 7%, you could withdraw the required amounts to meet exactly either the nonuniform series of expenses or pay out an equivalent amount of \$640. (Fig. B-1f)

Example B.10. Calculate present worth some years hence to present worth today.

Assume the expenses given in Example B.9 were to be associated with a piece of equipment to be installed 5 years from now. What is the present worth of the non-uniform expenses in that case? First, we refer to Figure B–1g. The present worth previously obtained was the present worth for the expenses incurred in the 10 years following installation of the project. This is a present worth five years from now. In terms of today's present worth, it is a future worth five years away. The present worth today is obtained simply by converting the future worth in 5 years to a present worth. From Table B–3, the future worth to present worth factor, 5 years, is 0.7130. Consequently:

Present worth today = \$4,496 × 0.7130 = \$3,206

Procedure for Economic Studies

The procedures for commencing an economic study may be laid out in a sequence of steps:

1. The facts concerning the different plans that could be used to meet the requirements of the problem should be set down. The plans should be made as comparable as possible.

2. The capital expenditures that will be incurred under each of the plans and the timing of these expenditures should be determined. The amounts and timing of operating and maintenance expenses must be estimated; allocations of cost to capital and expense must be adhered to.

3. A study period must be selected during which the revenue requirements incurred by the plans will be evaluated. In economic studies, it is seldom possible to find a study period that will precisely reflect the timing inherent in each of the plans under study. It will often be helpful to draw a diagram of the timing of capital and expense dollars for each of the plans in determining the study period. The study period chosen must be one determined on the basis of judgment. In every case, it must be sufficiently long to approximate the overall effects, over a long period of time, of the money reasonably to be spent for both capital and operating expenses.

4. The annual charges resulting from the capital expenditures in each phase must be calculated if broad annual charges cannot be applied. In considering alternate plans, items common to the several plans may be omitted from the calculations. The effect of temporary installations, salvage, and of the removal of equipment that can be used elsewhere on the system must be taken into account.

5. When annual revenue requirements are nonuniform, the present worth of the revenue requirements for each plan must be calculated. The most economical plan will have the lowest present worth of revenue requirements. In the case where annual revenue requirements are uniform throughout the study period, the plan with the lowest annual requirements will be the most economical.

6. The comparison of the economic differences among the plans may be made on the dollar differences among the present worth of the revenue requirements. If percentage difference is considered, the dollar differences may be misleading. In conducting the study, charges that are the same in the several plans are generally omitted. This will distort the base upon which a percentage difference is derived.

7. A recommendation of the most advantageous plan must be made. The plan with the minimum revenue requirements would be recommended from an economic point of view. Other considerations may indicate the recommendation of one of the other plans despite higher revenue requirements.

Conclusion

Economic studies constitute perhaps the most important ingredient in the implementation of a project. In sum, the consideration of any undertaking must answer satisfactorily three basic requirements or questions:

1. Why do it at all?
2. Why do it now?
3. Why do it this way?

The answers to these questions can, in large part, be supplied by the results of economic studies.

Appendix C

Meter Reading

The reading of customer meters—not only for electric service, but for water and gas services as well—is one of the most dynamic parts of the utility industry. A review of how this part of the business is changing—and the challenges these changes represent—is essential for any discussion about a total load management program.

Most utility revenue meter reading is currently performed manually. A utility representative visits a consumer location, reads the meter, records the information in some fashion, and returns the data to a central location. At this writing, only a relatively small portion of the millions of meters in the US are read by automatic devices that transmit data back to a central control, recording, and processing unit, a process known as *automatic meter reading* (AMR).

AMR is the remote collection of consumption data from customers' utility meters over telecommunications, radio, power-line and other links. AMR provides electric, gas, and water utilities with the opportunity to enhance service to customers; streamline metering, billing and collection activities, and gain a competitive advantage. AMR is the utility's gateway to present and future customers.

Meter Routes

Conventionally, a block of meters is set up for a meter reader to be able to cover in a normal working day in a relatively small geographic area.

These vary greatly with the characteristics of the area—in rural, mountainous, desert, or farm areas, the meters will have considerable distance between them compared to a suburban housing development of single-family homes, condos, or townhouses. Urban apartments, shopping centers, and other high-density locations also affect the meter route.

Billing Cycles

A common practice is to read meters every other month and estimate the bill in between using the *consumer use characteristic*.

Commercial and industrial meters are usually read every month because of their high energy use and demands, and because they usually represent a smaller percentage of the total meters on a system.

An estimated reading can be made by using the historical meter reading information, generally contained in computer files. This can be done for the different seasons and adjusted for any trend indicated by load additions or reductions from improving consumer energy efficiencies.

Utilities have also made available to residential consumers' budget payment plans, so they pay an estimated average fixed amount each month until the end of the budget period. At the end of the period the actual cost is reconciled with the total of the budget payments.

Hard-to-Read Meters

Inside meter locations present the most common impediment to physically reading a meter, because it depends on someone being there to let the meter reader in. This can make it costly to make an actual "read" if several visits are needed or special appointments have to be set up to get an actual reading. Complicating this is the fact that there may be, in some cases, a public service requirement to have an actual read within a certain time period. Some utilities utilize a remote meter dial mounted outside the house, which is connected to the inside meter register.

Other "hard-to-reads" are caused by physical obstructions and animals. The possible danger of physical harm to a meter reader inherent in some urban situations also comes under the hard-to-read category.

Cut-Ons and Cut-Offs

When ownership or tenant of a premise changes, the utility may remove the meter or discontinue the service temporarily to mark the change in consumer accounts.

Cut-ons are more usual for new construction, while a *cut-off* can be employed when a tenant or owner moves on. A cut-off can also be prompted by non-payment of a bill or theft of service. The latter requires a physical action in the field, either to actually change the service status by removing the meter or disconnecting the service wires at the pole and/or to read or inspect the meter.

Theft of Service. A residential meter can sometimes be removed and reinserted so that the register runs in reverse, or the meter removed and a jumper installed for a period of time to reduce the actual registration of energy used. To prevent this, most utilities use a meter seal having a unique registered number that is broken if the meter is removed for any reason. If an unauthorized meter removal were made, the meter reader or some other operating person would notice it. Programmed computer checks can also be made to show less-than-normal estimated usage and trigger an investigation. In remote automatic meter reading systems, the theft of service attempt would be detected at the time it occurs.

Metering, Billing, Customer Service, and Competition

Summary

Until now, such services have been provided by the local utility and included all of the administrative and operating functions associated with metering the amount of electricity used and billing individual customers for the generation, transmission, and distribution of electricity used. These services have been "bundled" together—that is, provided by one supplier.

Electricity restructuring plans undertaken by participating states thus far call for "unbundling" a utility's functions so that consumers can choose among different electricity generating companies. Some states are also considering "unbundling" metering and billing services, allowing them to be provided by various competing companies. This raises great many issues.

According to *The Chartwell AMR Report, 1997-1998, 3rd Edition*, a key issue in any restructured industry is whether utilities will own meters, or whether customers will be able to select their meters along with the electricity supplier. Will state regulators permit utilities to recover AMR installations costs as a part of any mandated changeover? How will electricity use be measured and costs allocated among various electricity suppliers? Who will set operating standards for electric meters? Who will install and maintain them? Who should have access to meter information and how should it be provided? How will bills be generated for various services provided by several suppliers? Who will customers call if they have a problem with billing or service?

One of the best studies on this subject is *Competitive Metering, Billing and Customer Services: An Analysis of Operational Issues, Part 1*, by Dr. Stephen S. George of Putnam, Hayes & Bartlett, Inc. His work is one study in a series of works examining retail electric competition issues prepared for the Edison Electric Institute (EEI).

Challenge 1: Measuring the Cost and Amount of Customer Energy Use

How much electricity costs relates to when it is generated. Electricity generated on a constant basis by large baseload power plants, supplying basic energy needed for a particular area, is usually the least expensive. To supplement it at peak-demand times, utilities purchase power from other utilities or bring on-line more expensive "peak-load" generating facilities. Determining the precise cost of a customer's electricity requires knowing when and how much power was consumed.

Under competition, customers who consume large amounts of power—manufacturing industries and large commercial businesses—want to be able to contract for electricity from a wide variety of producers who offer the lowest price on a consistent basis. To supplement their electricity meters—meters which provide information about electricity use on an hourly or even quarter-hourly basis—new systems will need to incorporate the increasingly complex electricity purchases from various and changing suppliers. Metering and billing services will have to track complex consumption patterns, integrate data coming from multiple sources, incorporate information about electricity purchases from different sources, and report that data to central billing facilities are required.

For residential and small business customers, meters indicating total consumption for a monthly billing cycle currently track electricity use. Such meters have been used for many years under a regulated electricity system, but were not always designed to show how much power an individual customer uses at specific times of the day. A competitive market may induce the development of such time-of-day meters—which have been used in some locations—but at the outset of competition, reliance will be made on *load profiling* to measure the electricity use of residential and small business customers.

Load profiling—estimating individual customer electricity use based on past consumption habits—can yield an approximate mix of baseload, peak-load, and supplementary other power for which a customer can be billed. As competition becomes the norm and customers switch among various electricity suppliers, approximations may become less accurate, leading to potential misallocation of costs among customers and companies selling electricity. A major challenge to utilities and energy suppliers will be to develop new and better ways to allocate costs under load profiling to ensure electricity customers and suppliers are charged accurately.

Challenge 2: Who Owns the Meters in a Competitive Market?

Meters have traditionally been the property of the utility supplying the service. The meter placed on a customer's property is a means of assessing the level of use and so generating an accurate billing. As competitive markets develop, however, the question will arise— who owns the electricity meter? The answer depends upon the types and terms of new services—and who will offer them. In addition, as new meters are developed to offer these new and different services, whose responsibility are they?

Consumers may imagine that one meter can serve the needs of all suppliers, but those who switch from one competitive supplier of electricity services to another may incur the added costs of purchasing a new meter. There may also be confusion for consumers over whom to call when problems arise. What happens if a utility crew is dispatched to investigate a power outage and discovers that problems exist with a meter owned by another service provider? Does this result in a second service call and a longer wait by the customer to have service restored?

Clear rules and procedures regarding competitive metering services will help bring benefits to all consumers and reduce confusion.

Challenge 3: Access to Meter Information

In competitive electricity markets, many different parties will require access to meter information. Procedures need to be put in place to allow an orderly flow from utility-only access to this greater demand.

- Consumers will want to know how much power they consume, and at what times, to help them control costs and electricity bills.
- Electricity Suppliers—including utilities, generating companies, and power marketers—will need access to meter data for customer billing, commercial settlements, bidding on electricity contracts for potential customers, and resolving disputes.
- Utility Distribution Companies (or DISCOS) will need access for customer billing, system operations, planning expansions or modifications to the distribution system, and dispute resolution.
- Government regulators will remain very involved in the transmission and distribution side of a competitive electricity business. Such policy makers will need access to consumption data to formulate consumer protection programs, adjudicate disputes, and to monitor market developments.

This goes far beyond political or "turf" considerations. Access to meter data by these parties raises a wide variety of issues associated with the physical act of meter reading, data management, and information dissemination. This includes:

- Standards governing the way computer chips in meters encode data and communicate it to computers
- Controlling who has access to meters and customer usage information
- Deciding what data to provide, and in what form
- Communicating data to all relevant parties
- Rules protecting customer privacy and controlling how the information is used

Challenge 4:
Billing and Collections

Billing and collections include several traditional functions. Under competition, a number of important issues related to these functions must be addressed. Traditionally, these functions include:

- Calculating charges for different services
- Developing standards for enrolling customers that protect consumers and suppliers
- Determining who should provide bills to the end-use customer
- Deciding what information should be in a customer's bill
- Collecting amounts due
- Establishing credit
- Resolving disputes among suppliers and customers
- terminating service.

What enrollment standards are necessary to protect consumers and suppliers?

How will providers control "slamming"—the unauthorized switching of customers to a service provider without the knowledge or approval of the customer?

Other concerns relate broadly to the kinds of information all retailers must convey to customers, including the characteristics of contracts. For example, regulators may require that customers be informed, in writing, of their rights to obtain service from a utility distribution company as a default provider.

Other concerns that have been suggested include the following:

- Who should be allowed to provide bills to the end-use customer? There are at least three options regarding how bills should be distributed: mandatory provision of a single bill by utility distribution customers; mandatory provision of a single bill by retailers; and allowing each entity to bill for the service it renders.

- What information should be provided on customer bills? Should bills include the costs of energy, transmission, ancillary services, utility distribution company charges, competition transition charges, state and local taxes? Should suppliers be obligated to delineate the type of generation being used? What about information overload for consumers who are satisfied with the cost of service and can get detailed information on request from their billing agent?
- How should disputes be handled among multiple suppliers? To best serve consumers, dispute resolution processes must be established that provide prompt redress, screen out complaints that stem from misunderstandings rather than abuse, maintain neutrality between customers and suppliers, and keep long-term records that allow examination of patterns over time.
- Who should be allowed to terminate service, and what procedures are needed to protect consumers and service providers? Currently, in most jurisdictions, only a utility distribution company is authorized to physically disconnect customers. Consequently, power suppliers are at a financial risk since the customer may continue to receive energy across the utility distribution company's wires even when payment is past due.

Challenge 5: Call Center Operations

Call centers are the point of contact between utilities and customers. They field customer questions, start and terminate service, arbitrate billing questions, disputes, service interruptions, and safety concerns, and help to market value-added service programs. As retail competition develops and distribution unbundling occurs, customer inquiries are likely to increase—yet the answers that call centers will offer will be increasingly complex and uncertain, at least at the outset.

Among the questions that must be answered are:

- What obligations do utility distribution companies and other suppliers have to respond to inquiries for services which they don't provide?
- What flexibility should utility distribution companies have to respond to customer inquiries for services offered by other suppliers?
- If metering and distribution services are separated, how can utility

distribution company customer information systems be integrated with third party systems to respond to even the most basic bill inquiry questions on a timely basis?

Challenge 6: The New Challenges in Technology

Changes in metering hardware—and in what it takes to read meters and to process the information gained—is an obvious development that will come from retail competition. Both EPRI and the Automatic Meter Reading Association (AMRA) has been studying these issues for a great while but, as noted in the other challenge areas, there is yet no consensus.

Utility company metering of electricity for billing purposes typically uses technologies that were developed a century ago and, as has been noted here, the overwhelming majority of meters are still read visually and manually recorded in a meter book by a person walking a route. New innovations such as the AIM System of American and others will enable utilities to totally redesign conventional revenue billing, load evaluation and survey, and to undertake customer relations programs necessary in an open-competition environment. Advanced meters include both expanded metering features and communications technologies:

- Time-of-use rates and demand profile recording
- Remote electronic meter reading
- Mobile radio meter reading
- Fully automatic meter reading (AMR) systems.

Handheld computers—with remote meter reading and remote electronic meter reading—have introduced automation into the meter reading and billing process, including access to hard-to-read meters. But these systems still require someone to physically contact each meter or a receptacle linked to a nearby meter.

Mobile radio systems reduce labor requirements and read errors by using low power radio to communicate directly between a meter and a handheld or van-based computer. Mobile radio systems have the largest number of installed units and have the highest acceptance by utilities. However, they are limited with regard to meter reading frequency (scheduled monthly reads only) and other advanced metering features.

For meter reading purposes, *dial-in system meters* are programmed to periodically call a central station, usually between midnight and 6 a.m., by using the existing telephone line. Ringing a customer's phone is not a part of this normal and usual meter-reading function. Information on customer usage is stored in a module until the scheduled call-in time. For two-way communication involving other more sophisticated functions, modification is necessary to telephone equipment to direct the utility's call to the customer without ringing his phone (see following *dial-out system*).

Conventional dial-out telephone systems with dedicated leased telephone lines is another form of an unrestricted full-time two-way communications AMR system for a limited number of special customers. Special customers include those with very hard-to-read meters or unique metering requirements (large customers). Some dedicated lines can be avoided by using automatic line-share switches, which connect a shared line to a meter without ringing the customer's telephone. Expansion to a large set of customers will require no-ring dial-out telephone systems. The no-ring dial-out telephone system is dependent on the cooperation of the telephone operating company for access, installation of special equipment at the telephone switch station, and reasonable tariffs. Telephone number management has proven to be a difficult problem for large systems.

Distribution line carrier systems perform well for high-density electric metering and DSM load control. These systems are independent of the telephone companies and radio frequency availability problems, but may be prohibitively expensive for very low density metering and may be unavailable for some gas and water metering.

Long range radio—configured as packet transceivers, cellular nodes or central broadcast towers—offer the largest range of near real-time meter reading for large metering systems. This is the only technology that can provide unrestricted high-speed two-way communication with each individual meter of a large system, which may be necessary for additional services such as security, energy management or customer information. However, the radio equipment required for complete coverage of a large service territory, and the maintenance thereof, is the most expensive of all the systems.

Unfortunately, what EPRI and the AMRA (among others) have been able to determine in continually studying the characteristics and relative merits of all the major AMR system technologies, is that each technology excels in one or more areas and each also has a technical weakness or high cost. In short, there is no single "best" AMR system for all applications. Selection of the most appropriate AMR system will require investigation of the characteristics of the service territory to be metered, the number and type of customers to be metered, and the metering services required.

Utility Metering

Basic electric, gas and water utility meters, their capabilities, and some of the advanced metering equipment that is commercially available are summarized below. Many advanced metering functions are currently available through add-on equipment and multiple systems. Advanced metering equipment will allow consolidation of these functions into one common system, which can reduce equipment and operating costs.

Electric Meters

Electric power is typically measured using the standard kilowatt hour (kWh) electric power meter, which is found on about 100 million residences and commercial buildings in the United States. Utility companies purchase these meters for $20 to $30 each. The standard spinning disk kWh meter technology was developed a hundred years ago. And, as was the practice at its first commercial introduction, the overwhelming majority of these meters are still read visually and manually recorded in a meter book by a person that walks a route.

The basic kWh meter measures accumulated electrical energy (kilowatt-hours). The kWh is of prime importance since it relates directly to the amount of electricity delivered to the customer, and therefore, billed to them. Although many large commercial customers are billed for kilowatt (kW) demand (maximum power) and power factor in addition to accumulated energy (kWh), that is rarely true for residential customers. Some electric utility rate structures have introduced time-of-use (TOU) rates, which charge higher rates during peak demand times than during off peak times. The basic kWh meter is capable of measuring and displaying accumulated energy (kWh), but not instantaneous power or time-of-day use.

A modified basic kWh meter can be used to measure, and record, demand interval power. A timer is included to set the duration of the demand interval. Such a meter always integrates demand over a pre-set time interval (15, 30 or 60 minutes) and does not record true instantaneous demand.

Electric utilities also charge some large commercial and industrial customers according to power factor (PF). Measurement of PF requires the use of a basic meter that has been modified to measure reactive power instead of kWh.

Modified Spinning Disk Meters

The first step toward advanced electric meters is based on a standard spinning disk kWh meter combined with some type of pulse initiator and/or encoding device, resulting in an electro-mechanical meter. Solid state electronics are used for monitoring and storing metering data over extended periods of time (60 days of more). These meters can store large amounts of data, including both billing (total accumulated kWh, 15, 30 or 60 minute demand, and TOU) and load survey (1 minute to daily interval profiles) information. These meters usually have an industry standard optical communications ports (typically compatible with OpticomTM) and/or RS232 computer communications ports. The most advanced electromechanical meter systems include sophisticated communications capabilities, such as power line carrier, telephone modems, or radio transceivers, and still fit under the glass of a standard kWh meter. Use of the appropriate interface device and hand-held computer enable a meter reader to download large amounts of data from the meter directly into the hand-held. Additionally, many of these meters can be reprogrammed by means of the same hand-held computer. At a retrofit cost of $100 to $250 per meter, this may be the least expensive electronic metering option for existing meters.

Solid-State Electronic Meters

The second step toward advanced electric meters completely replaces the mechanical components of standard spinning disk kWh meters with solid-state electronics. Such meters were first introduced in the 1970s. Most use transducers to convert the service current and voltage into low voltage analog signals and/or pulses. Electronic circuits convert the analog signals into digital values, which are used to calculate a variety of electrical parameters, such as kWh, Watt, PF, Volt, Amp, etc. The solid-state electric power meter is capable of measuring much more than a retrofitted standard mechanical meter. Solid state meters have become quite sophisticated, and can measure and store huge amounts of selected electrical parameters, load survey (1 minute to daily interval profiles) information about any of electrical parameters, TOU data, total accumulated values, etc.

Like the electromechanical meters, solid state meters usually have industry standard optical communications ports (typically compatible with Opticom™) and/or RS232 computer communications ports. Use of the appropriate interface device and hand-held computer enable a meter reader to download large amounts of data from the meter directly into the hand-held. The meters can be reprogrammed by means of the same hand-held computer. In addition, most of these meters include Supervisory Control and Data Acquisition (SCADA), Distribution Automation (DA) or AMR communications equipment, such as a radio transponder, power line carrier or telephone modem circuit. These meters tend to be used in substations and on particularly important industrial loads, and as such are quite useful. But, at $250 to $1500 each, they are too expensive for most mass AMR applications.

Appendix D

Electric Utility Organization and Operation

Electric utilities can be classified by size, type of operation, ownership, and territory served. While there are fewer "typical" or "usual" organizations in the current climate of competition and corporate "re-engineering" evident in the past few years, there are some generalizations that can be made.

It should also be noted here that an electric utility may also be in the natural gas business (or be owned by a natural gas company!) or other commercial endeavors. As the natural gas and electric power industries continue to combine—a term commonly used is *convergence* into a *Btu stream*—this element will become more and more important.

Utility Organization

There are broad general function splits between support services and actual operations in the medium-to-large size utilities that may not be found in smaller utilities (see Table D–1).

Table D–1 Utility Support Services and Operations

Support Services	Operations
legal	generation
finance	system operation
accounting	line construction
public relations	line maintenance
human resources	distribution operation
purchasing	major construction
computer services	emergency (storm) restoration
transportation	
engineering and survey	
load research and rates	

Services Found in Either Category
system planning
meters
meter reading
stores
distribution engineering
mapping
general shops
consumer facility engineering

If a utility is large enough, divisions of functions and areas may be established that then report back to a central management group or authority.

Utility Classification—Ownership

Investor-Owned

Investor-owned electric utilities (IOUs) can be organized as either individual corporations or holding companies. The more prevalent of the two organizations is the individual corporation. Under a holding company organization, a parent company is established to own one or more operating utility companies that are integrated with another. In an individual corporation, the utility is a so-called "operating company." This means that the utility provides basic services for the generation, transmission, and distribution of electric power to consumers. The majority of all IOUs perform all three functions.

Investor-owned or private utilities can be found in all sizes and in all areas of the United States. Stockholders who expect a return on their investment own these utilities. IOUs operate in all states except for Nebraska, where the utilities primarily consist of municipal systems and public power districts.

Most IOUs sell power at retail rates to several different classes of customers and at wholesale rates to other utilities—including other IOUs—as well as federal, state, and local government entities, public utility districts, and rural electric cooperatives. In 1995, only 244 of the 3,199 U.S. utilities were IOUs, but accounted for more than 75% of both utility sales to consumes and total utility generation.

Like all private businesses, the fundamental objective of an IOU is to produce a return for investors. The IOU either distributes its profits to stockholders as dividends or reinvests these profits. It is granted a service monopoly in certain geographic areas and is obliged to serve all consumers. As franchise monopolies, IOUs are regulated and are required to charge reasonable prices, to charge comparable prices to similar classifications of consumers, and to give consumers access to services under similar conditions. Such utility corporations are not guaranteed a profit but do have a maximum return on their investment set by a public authority, who determines the rate tariffs charged to consumers. These utilities pay both income taxes and property taxes. The largest investor-owned utility in Texas is Texas Utilities Company, Dallas, with a summer peak load of 18,000 MW (1990) serving a population of 5,000,000.

It must be emphasized that in the current push to deregulate and thus make competitive the U.S. electric utility industry, this historic structure of the IOU sector (as well as others) is likely to change.

Cooperatively Owned Utilities

A second classification of ownership is the *Rural Electric Membership Corporation (REMC)* set up by the Federal Rural Electrification Administration (REA). These were originally set up to bring electric energy to the more rural areas of the country but now also serve fair-sized towns. They frequently will have the designation "cooperative" or "association" associated with their geographic name.

Most rural electric cooperative utilities are formed and owned by groups of residents in rural areas to supply power to those areas. Some cooperatives may be owned by a number of other cooperatives. There are three types of cooperatives: distribution only, distribution with power supply, and generation and transmission.

Cooperatives operate in 47 states and they represent 29% of the total number of utilities in the country. Most distribution cooperatives resemble municipal utilities in that they often do not generate electricity, but purchase it from other utilities. Generation and transmission cooperatives are usually referred to as power supply cooperatives and are normally owned by the distribution cooperatives to whom they supply wholesale power. Distribution cooperatives resemble federal utilities, supplying electricity to other utility consumers from their generating capability. Cooperatives accounted for 6% of total utility generation and 8% of utility sales to consumers in 1995.

Publicly Owned Utilities

As stated earlier, most, but not all, electric utilities are generators of electric power. Publicly owned electric utilities are such an example. They can be divided into two different categories: those that do generate power, and those that do not generate power. To recap, generators are those electric utilities that own and operate generating capacity to supply some or all of their customers' needs. However, some generators supplement their production by purchasing power. The nongenerators rely exclusively on power purchases. The nongenerators comprise over half of the total number of publicly owned electric utilities.

In 1995, publicly owned electric utilities accounted for nearly 63% of the number of electric utilities in existence in the United States. They produced 11% of total utility generation and accounted for 14% of utility sales to consumers. Other publicly owned electric utilities include:

- Municipal authorities
- State authorities
- Public power districts
- Irrigation districts, and
- Other state organizations

Municipal utilities tend to be concentrated in cities where the loads are small. They exist in every state (except Hawaii), but most are located in the Midwest and Southeast. The political area served usually owns the municipals. As of this writing, they do not pay taxes as investor-owned utilities do and frequently enjoy lower cost of financing because of their public ownership and ability to raise money through taxes on their constituents. Most municipals are small in size, although some urban "munis" are large in scale, such as the Austin Electric Utility Department with a summer peak load of 1,483 MW (1990) and 265,000 consumers.

State authorities are utilities that function in a similar manner to federal utilities, generating or purchasing electricity from other utilities and marketing large quantities in the wholesale market to groups of utilities within their states at lower prices than the individual utilities would otherwise pay. Publicly owned utilities are mainly located in Nebraska, Washington, Oregon, Arizona, and California. In general, publicly owned utilities tend to have lower costs than IOUs because they often have access to tax-free financing and do not pay certain taxes or dividends (payments to shareholders based on utility returns). Publics also have high-density service areas.

Government Owned Utilities

The fourth class of ownership consists of *federal, state, and district systems*. These are similar to "munis" in that they pay no taxes and have very low rates of financing. They tend to cover large geographic areas and are generally wholesalers to other electric utilities. Well-known examples are the Tennessee Valley Authority (TVA), Bonneville Power Authority (BPA), and the New York Power Authority (NYPA). A state-owned Texas example is the Lower Colorado River Authority, operating out of Austin with a net peak load of 2,257 MW in 1997.

Currently, there are 10 federal electric utilities in the United States. According to the government agency, Energy Information Administration, they are:

- The Department of Energy's five power marketing administrations (Alaska, Bonneville, Southeastern, Southwestern, and Western Area Power Administrations)
- The Department of Defense's U.S. Army Corps of Engineers
- The Department of the Interior's U.S. Bureau of Indian Affairs

- The Department of the Interior's U.S. Bureau of Reclamation
- The Department of State's International Water and Boundary Commission
- The Tennessee Valley Authority

Of these 10 utilities, three are major electricity producers. They are the Tennessee Valley Authority or TVA, the U.S. Army Corps of Engineers or USCE, and the U.S. Bureau of Reclamation or USBR.

TVA is the largest federal power producer, marketing its own power in both the wholesale and retail markets. Generation by the USCE, except for the North Central Division (Saint Mary's Falls at Sault Ste. Marie, Michigan), and by the USBR is marketed by four of the federal power marketing administrations or PMAs: Bonneville, Southeastern, Southwestern, and Western Area. These four PMAs also purchase energy for resale to other electric utilities in the United States and Canada. Alaska, the fifth PMA, operates its own power plants and distributes power to consumers.

In 1995, federal power plants generated about 9% of total utility electricity in the United States (this percentage appears to hold steady year-after-year). The power is generated mostly from hydroelectric (or water-driven) facilities. TVA generates electricity from coal and nuclear power as well as from hydropower. In 1995, only less than 2% of the utility generation was sold by federal utilities to retail consumers. Consumers of federal power are usually large industrial customers or federal installations (i.e., military bases, etc.). Most of the remaining energy produced by federal utilities is sold in the wholesale market to publicly owned utilities and rural electric cooperatives for resale. These wholesale consumers have preference claims to federal electricity. Only the remaining surplus, after meeting the energy requirements of preference consumers, is sold to IOUs.

Power Pools

Overlaying the individual utilities are operating entities known as *"power pools"* that coordinate interchange of power between individual companies and between other power pools. Regional groups of power pools are often organized as coordinating councils, such as the Middle Atlantic Coordinating Council (MACC). These regional councils, as well as the individual power pools, establish area transmission and generation plans, operating standards, and pricing of energy. There is also environmental involvement through committees made up of representatives of the individual companies.

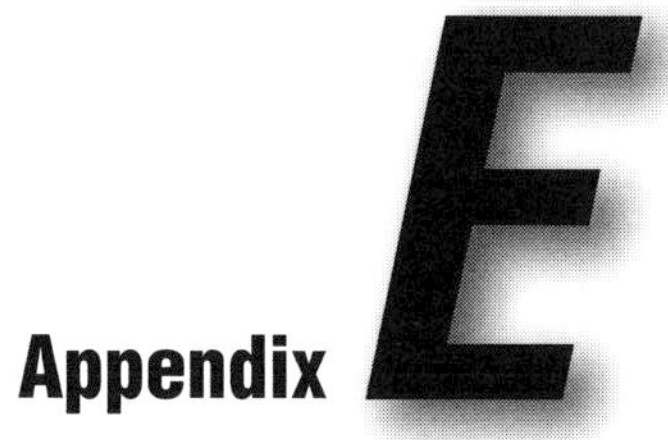

Electric Power Glossary

Accuracy. The extent to which a given measurement agrees with a standard defined value. For revenue metering, public service authorities require stringent, very high degrees of accuracy in watt-hour and demand meters. These include pre-installation testing and calibration as well as periodic testing, which may be done on a sampling basis for electric utilities. Remote meter reading devices must also meet both the utility standard and public service authority standards for accuracy.

Adjustment clauses. A clause or series of clauses in a rate schedule (tariff) that permits a utility to charge for changes in fuel costs, temperature deviations from normal, cost of purchased power, ratchet demands, etc.

Allocation. The procedural step in a cost of service study whereby joint costs are allocated among consumer classes based on demand, energy, or some other cost-related feature of service.

Allowable costs. Expenses that are allowed as operating costs chargeable to the consumers; certain other costs are chargeable to the stockholders or other owners.

Ammeter. An instrument to measure current flow, usually indicating in amperes. Where current is measured in milliamperes (1/1,000 of an ampere), the instrument may be called a milliammeter.

Ampere (amp). The unit of measurement of electric current. It is proportional to the quantity of electrons flowing through a conductor past a given point in 1 second. It is the unit current produced in a circuit by 1 volt acting across a resistance of 1 ohm.

ASE cable. A variant of SE cable in which a flat, steel strip is inserted between the neutral conductor and the outside braid for greater mechanical strength.

Automatic meter reading. A method of reading a meter (watt-hour, demand, gas, water, or any other type of meter), preparing and conditioning the data, and transmitting the accumulated information from the meter location back to a central data accumulation device. This central collection device in most cases is some form of computer. The communication link may be radio, telephone line, power line carrier, cablevision, or any combination thereof.

Average annual electric bill. Annual electric revenue from a class of service divided by the average number of such consumers for the 12-month period.

Average cost. A method of determining the cost of providing service to the various consumer classes. Average cost-of-service figures may be used in setting rates. Average costs are total costs divided by the number of units produced. This method, while distinguishing costs between different consumer classes, fails to recognize that not all kilowatts and kilowatt-hours are produced at the same cost within one consumer class. Seasonal, time-of-day, and marginal cost-based rates more accurately reflect the true costs of producing each kilowatt or kilowatt-hour.

Avoided costs. The costs an electric utility would otherwise incur to generate power if it did not purchase power from another source.

Basic reference standards. Those standards with which the value of electrical units are maintained in a laboratory, and which serve as the starting point of the chain of sequential measurements carried out in the laboratory.

Bottom-connected meter. A meter having a bottom-connection terminal assembly. Also referred to as an A-base meter.

Capability. In general, the maximum load in amperes, kW, or kVA that a system or a component of a system can carry without exceeding its design limits. They are usually defined as being normal or emergency limits. Normal limits are those that can meet expected conditions in the normal operation of the system without damage to the facility or without incurring a significant loss of expected life of the item. Emergency capability is some level above the normal limit that takes into account specific conditions of ambient temperature, preceding loading, and a calculated loss of life of the equipment.

Capacitance. That property of an electric circuit that allows storage of energy and exists whenever two conductors are in close proximity but separated by an insulation or dielectric material. When direct voltage is impressed on the conductors, a current flows momentarily while energy is being stored in the dielectric material. It stops when electrical equilibrium is reached. With an alternating voltage between the conductors, the capacitive energy is transferred to and from the dielectric material, resulting in an alternating current flow in the circuit.

Central station. Control equipment, typically a computer system, which can communicate with metering and load-control devices. The equipment may also interpret and process data, accept data from other sources, and prepare reports or consumer bills.

Class designation. The maximum of the watt-hour meter load range in amperes.

Connection charge. An amount to be paid by the consumer in a lump sum, or in installments, for connecting a consumer's facilities to the electric system.

Constant kilowatt hour. Pertaining to a meter (register constant, dial constant): the multiplier applied to the register reading to obtain kilowatt-hours.

Creep. For mechanical meters, a continuous motion of the rotor of the meter with normal operating voltage applied and the load terminals open-circuited. For electronic meters, a continuous accumulation of data in a consumption register when no power is being consumed.

Demand. The rate at which electric energy is delivered to or by a system, part of a system, or a piece of equipment. It is expressed in kilowatts, kilovolt-amperes, or other suitable unit at a given instant or averaged over any designated period of time. The primary source of demand is the power-consuming equipment and devices of consumers.

Demand charge. That portion of the charge for electric service based on the peak load furnished within a time period according to the established tariff.

Demand constant. Pertaining to pulse receivers, the value of the measured quantity for each received pulse, divided by the demand interval, expressed in kilowatts per pulse, kilovars per pulse, or other suitable units. The demand interval must be expressed in parts of an hour such as 1/4 for a 15-minute interval of 1-12th for a 5-minute interval.

Demand interval. The period of time during which the electric energy flow is averaged in determining demand, such as 60 minutes, 30 minutes, 15 minutes, or instantaneously.

Demand meter. A metering device that indicates or records the demand and/or the maximum demand, or both. Since demand involves both an electrical factor and a time factor, mechanisms responsive to each of these factors are required, as well as an indicating or recording mechanism. These mechanisms may be either separate from or structurally combined with one another. An alternative mode would be to have a computer interpret and calculate the desired demand (see integrated demand).

Demand register. A mechanism (for use with an integrating electricity meter) that indicates maximum demand and also registers energy (or other integrated quantity).

Demand side management (DSM). The broad term for electric load management applying to utility actions, programs, and designs for the purpose of lowering system peaks and reducing energy consumption by the consumer as well as the utility system. There are direct actions and controls by the utility that result in immediate results to lower peaks and reduce energy requirements. There are also indirect (passive) programs requiring the cooperation and participation by the consumer to achieve similar results.

Detent. A device installed in a meter to prevent reverse rotation.

Dial-out capability. The ability of a meter to initiate communications with a central station, usually using telephone lines.

Disk constant. *See* watt-hour constant.

Disk position indicator. Also known as a "caterpillar," it is an indicator on the display of a solid state register that simulates rotation of a disk at a rate proportional to power.

Diversity. That characteristic of a variety of electric loads whereby individual maximum demands usually occur at different times. This permits design of capability to meet a demand reduced from the sum of all the individual peak demands.

Dump energy. Energy generated that cannot be stored or conserved when such energy is beyond the immediate use in a system. This energy is usually bid out to neighboring systems at a price less than the cost to produce it.

Electromagnet. A magnet in which the magnetic field is produced by an electric current. A common form of electromagnet is a coil of wire wound on a laminated iron core, such as the voltage coil of a watt-hour meter stator.

Embedded coil. A coil in close proximity to, and nested within, a current loop of a meter to measure the strength of a magnetic field and develop a voltage proportional to the flow of current.

Encoder. A device that converts a meter reading into a form suitable for communicating to a remote central location or to a portable recording device or remote dial.

Energy charge. That portion of the charge for electric service based on the energy (kWh) consumed by the consumer.

Energy conservation. The strategy of a utility leading to programs for the reduction of electric energy consumption by the consumer. This includes rebates and/or low-cost financing assistance for the consumer to install or replace appliances, lighting, and motors with more efficient ones. It also includes using higher grades of insulation to reduce heat loss in cool weather and to reduce heat absorption in warm weather. Utilities will often furnish information and technical assistance at no cost to the consumer, but it is the consumer who will make the decision to increase his capital cost to reduce his energy costs. In addition to reducing energy consumption, conservation also lowers the peak loads.

Functional accounts. Groupings of plant and expense accounts according to the specific function in the electric system, sometimes referred to as the Uniform System of Accounts. For instance, there would be an account number such as 13804 or some other specific number for the capital value of all the meters a utility owns.

Gear ratio. The number of revolutions of the rotating element of a meter for one revolution of the first dial pointer.

Grounding conductor. A conductor used to connect any equipment device or wiring system with a grounding electrode or ground system.

Incentive rate. Some form of reduced rate generally designed to provide an incentive for targeted consumers to remain in the service territory or to lure other businesses to the territory, usually offered for a fixed period of time.

Time-of-day rates are in some sense an incentive rate to the consumer to time his load use so as to minimize his energy or demand use during peak load periods.

Inductance. That property of an electric circuit that opposes any change of current direction through the circuit. In a direct current circuit, where current does not change in direction, there is no inductive effect. In alternating current (AC) circuits, the current is constantly changing direction, so the inductive effect is appreciable. Changing current produces changing flux, which, in turn, produces induced voltage. The induced voltage opposes the change in applied voltage, hence the opposition to the change in current. Since the current changes more rapidly with increasing frequency, the inductive effect also changes with frequency.

Integrated demand. The summation of energy units through a definite period of time called the demand interval. (*see* **demand interval**)

Instrument transformer. A transformer that reproduces in its secondary circuit, in a definite and known proportion, the voltage or current of its primary circuit, with the phase relationship substantially preserved.

IR losses. Current x resistance.

K_e. *See* **KYZ output constant.**

K_h. *See* **watt-hour constant.**

K_m. *See* **mass memory constant.**

kVA. One thousand volt-amperes

kW. One thousand watts.

kWh. One thousand watt-hours.

KYZ output constant (K_e). The pulse constant for the KYZ outputs of a solid state meter, programmable in unit-hours per pulse.

Lagging current. An alternating current that, in each half cycle, reaches its maximum value a fraction of a cycle later than the maximum value of the voltage that produces it.

Leading current. An alternating current that, in each half cycle, reaches its maximum value a fraction of a cycle sooner than the maximum value of the voltage that produces it.

Load compensation. That portion of the design of a watt-hour meter that provides good performance and accuracy over a wide range of loads. In modern, self-contained meters, this load range extends from load currents under 10% of the rated meter test amperes to 667% of the test amperes for class 200 meters.

Load control. A direct action by the utility to shift load off-peak, either by a programmed computer command or by manual implementation of remote control. Voltage reduction and cycling of motor and appliance loads and water/house heating are examples of load control.

Load curve. A curve on a chart showing energy, power, or amperes plotted against time of occurrence to show the varying magnitude of a load during the period covered.

Load factor. The ratio of the average load in kilowatts (kW) supplied during a designated period to the peak or maximum load in kW occurring in that period. Load factor, in percent, also may be derived by multiplying the kilowatt-hours in the period by 100 and dividing by the product of the maximum demand in kW and the number of hours in the period.

Load forecast. A predicted demand or energy amount expected during a period of time or at a specific instant in time. Load forecasts may be short-term for operating purposes, long-term for system planning purposes, or any range in between. Forecasts may be of total system or regional load, or of areas of the system such as served by a substation or distribution circuit. They might also be of consumer loads (especially in the case of transmission consumers), or appliance/device loads, including street lighting.

Load research. Generally refers to a utility activity designed and carried out to determine consumer load characteristics. The results of load research activities are not only used in rate analysis and development but also to determine the electric design parameters, both for the system and the components of the system.

Load shape control. *See* load shifting.

Load shifting. Reducing the system peaks by curtailing load directly or by a rate structure that motivates the consumer to move some of his energy requirements from peak load hours to off-peak usage. Filling the dips in the load curve improves the system load factor and makes more use of lower cost energy produced by base load units or off-peak purchased power.

Load survey. The measurement of the electrical characteristics of a consumer or segment of the electric system. This is usually done by portable special load monitoring instruments installed either on the consumer premises or on some part of the electric supply system to measure demand, energy usage, amperes, voltages, and/or power factor. An alternative to special field instrumentation would be to transmit consumer data back to a central location and develop the required information through computer programs.

Loss compensation. A means for correcting the reading of a meter when the metering point and point of service are physically separated, resulting in measurable losses. These losses include IR losses in conductors and transformers, and iron core losses. These losses may be added to, or subtracted from, the meter registration.

Loss factor. The ratio of the average loss in kilowatts during a designated period to the peak, or maximum loss of kilowatts occurring in that period.

Losses. The general term applied to energy (kilowatt-hours) and power (kilowatts) lost in the operation of an electric system. Losses occur principally as energy transformations from kilowatt-hours to waste heat in electrical conductors and equipment or devices.

Mass memory constant (K_m). The value, in unit quantities, of one increment (pulse period) of stored serial data. For example:

$$K_m = 2{,}500 \text{ W-hr/pulse.}$$

Maximum demand. The demand usually determined by an integrating demand meter or by the integration of a load curve. It is the summation of the continuously varying instantaneous demands during a specified time interval. With the computer technology available, the integration of demand can also be accomplished through automatic meter readings transmitted from the consumer meter back to a central location. This makes it possible not to have a demand meter or register at the location of usage.

Memory. Electronic devices that store instructions and data. Volatile memories can be written to, and read from, repeatedly. Random-access memories (RAM) require uninterrupted power to retain their contents. Read only memories (ROMs) are programmed once, may only be read (but can be read repeatedly), and do not require constant power to retain their contents. ROMs are typically used to store firmware in dedicated systems.

Ohm. The practical unit of electrical resistance. It is the resistance that allows 1 ampere to flow when the impressed electrical pressure is 1 volt.

On-peak demand register. A register that will record the total energy used and, in addition, will register maximum demand during on-peak periods. The control of the demand recording is with a solenoid-operated demand gear train that may be actuated from a local device or from a device located remotely.

Percent registration. Percent registration of a meter is the ratio of the actual registration of the meter to the true value of the quantity measured in a given time, expressed as a percentage. Percent registration is also sometimes referred to as the accuracy of the meter.

Phantom load. A device that supplies the various load currents for meter testing, used in a portable form for field testing. The power source is usually the service voltage, which is transformed to a low value. The load currents are obtained by suitable resistors switched in series with the isolated low-voltage secondary and output terminals. The same principle is used in most meter test boards.

Phase angle. The phase angle or phase difference between a sinusoidal voltage and a sinusoidal current is defined as the number of electrical degrees between the beginning of the cycle of voltage and the beginning of the cycle of current.

Photoelectric meter. Also known as a counter, this device is used in the shop testing of meters to compare the revolutions of a watt-hour meter standard with a meter under test. The device receives pulses from a photoelectric pickup, which is actuated by the anticreep holes in the meter disk (or the black spots on the disk). These pulses are used to control the standard meter revolutions on an accuracy indicator by means of various relay and electronic circuits.

Power factor. The ratio of real power (kW) to apparent power (kVA) at any given point and time in an electrical circuit. Generally it is expressed as a percentage ratio. Resistance-type loads, such as incandescent lamps or electric resistance heating, are characterized as 100% PF loads. Electric motor loads in refrigerators and air conditioners are around 80% PF or lower.

Primary/transmission metering. When consumer loads reach a magnitude such that they will benefit from higher service voltage rates, the revenue will be metered by special metering installations called primary metering. This means the consumer is served at a distribution voltage of 4,000 volts (V) or higher (usually 13,000 V) or by a transmission line, usually operating at 69,000 V or higher. In these cases, the consumer furnishes his own step-down transformers and high-voltage service equipment. The utilities use potential transformers (PTs) and current transformers (CTs) to reduce both potential and current to levels for which the meter is designed.

Pulse device. For electric metering, it is the functional unit for initiating, transmitting, retransmitting, or receiving electric pulses, representing finite quantities, such as energy. Normally pulses are transmitted from some form of electric meter to a receiver unit.

Pulse initiator. Any device, mechanical or electrical, used with a meter to initiate pulses, the number of which is proportional to the quantity being measured. It may include an external amplifier or auxiliary relay or both.

Q-hour meter. An electricity meter that measures the quantity being obtained by lagging the applied voltage to a watt-hour meter by 60°, or for electronic meters by delaying the digitized voltage samples by a time equivalent to 60° (electrical).

Rate base. The value established by a regulatory authority, upon which a utility is permitted to earn a specified rate of return. Generally this represents the amount of property used and useful in servicing of consumers.

Reactance. The measurement of opposition to the current flow in an electric circuit caused by the circuit properties of inductance and capacitance, normally measured in ohms.

Reactive power. The portion of apparent power that does no work. It is commercially measured in kilovars (volt-amperes reactive). Reactive power must be supplied to most types of magnetic equipment, such as motors. It is supplied by generators or by electrostatic equipment known as capacitors.

Real power. The energy or work-producing part of apparent power; the rate of supply of energy, measured commercially in kilowatts. The product of real power and length of time is energy, measured by watt-hour meters and expressed in kilowatt hours (kWh).

Register constant. The number by which the register reading is multiplied to obtain kilowatt-hours. The register constant on a particular meter is directly proportional to the register ratio, so any change in ratio will change the register content.

Register freeze. The function of a meter or register to make a copy of its data, and perhaps reset its demand, at a pre-programmed time after a certain event (such as demand reset) or upon receipt of an external signal. It is also known as self-read, auto-read, or data copy.

Register ratio. The number of revolutions of the gear meshing with the worm or pinion on the rotating element for one revolution of the first dial pointer.

Registration. The registration of the meter is equal to the product of the register reading and the register constant. The registration during a given period of time is equal to the product of the register constant and the difference between the register readings at the beginning and the end of the period.

SE cable. Service entrance cable usually consists of at least two conductors with appropriate insulation, laid together and covered with a jacket material about which is wrapped a stranded, bare neutral conductor. An outer covering over the neutral conductor is a flame-retarding and waterproof braid. (*See* also **ASE cable.**)

Service entrance conductors. For an overhead service, that portion of the service conductors that connects the service drop to the service equipment. The service entrance for an underground service is that portion of the service conductors between a terminal box (external or internal) and the service equipment. In the absence of a terminal box, the service entrance runs from the point of entrance into the building to the service equipment junction.

Socket. Also known as a trough, it is the mounting device consisting of jaws, connectors, and enclosure for socket-type meters. A mounting device may be either a single socket or a trough. The socket may have a cast or formed enclosure, the trough an assembled enclosure which may be extended to accommodate more than one mounting unit, as may be encountered in multiple occupancy buildings.

Stator. The unit that provides the driving torque in a watt-hour meter. It contains a voltage coil, one or more current coils, and the necessary steel to establish the required magnetic paths.

Submetering. The metering of individual loads within a building or subdivisions of property (such as a trailer park) for billing purposes by the owner. For that application, usually there is one master meter by the utility for billing the owner and the sub-meters are used by the owner to charge tenants for energy used.

Temperature compensation. In reference to a watt-hour meter, refers to the factors included in the design and construction of a meter that make it perform with good accuracy in a wide range of temperatures. In modern electric meters this may range from –20° F to 140° F. In a gas meter installation, this temperature compensation may also mean the factor applied to compensate for the difference in temperature between inside metering and outside metering.

Three-rate watt-hour meter. A watt-hour meter with three sets of registers. It is constructed so that the off-peak energy will be recorded on one set of dials, and the on-peak energy for two different on-peak periods will be recorded on the other two sets of dials. The control of the recording system is by an internal switch or remote control system. In the event of a power failure, carryover for an internal time switch can be accomplished by battery or spring storage.

Time division multiplication. An electronic measuring technique that produces an output signal proportional to two inputs, for example, voltage and current. The width or duration of the output signal is proportional to one of the input quantities; the height is proportional to the other. The area of the signal is then proportional to the product of the two inputs.

Time-of-day rates. A rate strategy enacted by the utility to influence the consumer to change his load pattern and shift loads to off-peak hours by charging a higher rate for usage during peak hours and lower rates for off-peak hours.

Time-of-use metering. A metering method that records consumption and demand during selected periods of time. This allows consumption and demand (in some cases for commercial loads or very large houses) to be billed at different rates established in the utility tariff.

Transducer. A solid-state electronic device to receive energy from one system and supply energy, of either the same or different kind, to another system. This occurs in such a manner that the desired characteristics of the energy input appear at the output. For instance, it may take an AC voltage frequency or magnitude from a suitable interface and transmit it to a reading device or control relay.

Trough. *See* **socket.**

Two-rate watt-hour meter. A watt-hour meter with two registers or sets of dials, constructed so that the off-peak energy will be recorded on one set of dials and the on-peak energy on the other set. The control of the recording system is by an internal switch or external remote control signal. In the event of a power failure, carryover for an internal time switch can be accomplished by battery or spring storage.

V. *See* **volt.**

VA. *See* **volt-ampere.**

Var-hour meter. An electricity meter that measures and registers the integral, with respect to time, of the reactive power of the circuit in which it is connected. The unit in which this integral is measured is usually the kilovar-hour. This is most often used to obtain power factors.

Volt. The unit of electromotive force or electric pressure. It is the electromotive force that, if steadily applied to a circuit having a resistance of 1 ohm, will produce a current of 1 ampere; abbreviated V.

Voltage control. A utility's actions to reduce peak load, either on a planned basis or in an emergency situation in which generation or tie capacity is not available to meet the demand. A signal from a system operator to a remote device at substations automatically reduces the outgoing primary voltage and results in a temporary immediate load reduction.

Voltage of a circuit. The electric pressure of a circuit or system measured in volts. It is generally a nominal rating based on the maximum normal effective difference of potential between any two conductors of a circuit. In a typical house service and meter, this nominal voltage of 120 volts (V) can fluctuate between 117 V and 124 V due to consumer load variations.

Volt-ampere. The basic unit of apparent power. The volt-amperes (VA) of an electric circuit is the mathematical product of the volts and amperes of the circuit. The practical unit of apparent power is the kilovolt-ampere (kVA), which is 1,000 volt-amperes. An average residential service may have an apparent power of 5–10 kVA compared to a primary distribution circuit capability of 8,000 kVA and a transmission circuit capability of 100,000–700,000 kVA.

Watt. The electrical unit of real power or rate of doing work. The rate of energy transfer flowing due to an electrical pressure of 1 volt at unity power factor. One watt is equivalent to approximately 1/746 horsepower, or 1 joule per second. An average size incandescent light bulb uses 75 watts.

Watt-hour. The total amount of energy used in 1 hour by a device that requires 1 watt of power for continuous operation. Electric energy is commonly sold by the kilowatt-hour (kWh; 1,000 watt-hours).

Watt-hour constant (K_h). For an electromechanical meter, the number of watt-hours represented by one revolution of the disk. It is determined by the design of the meter and not normally changed; also called the disk constant. For a solid state meter (K_h or K_t): the number of watt-hours represented by one increment (pulse period) of serial data. Example: K_h or K_t = 1.8 watt-hours/pulse.

Watt-hour meter. An electricity meter that measures and registers the integral, with respect to time, of the active power (often referred to as real power) of the circuit in which it is connected. This power integral is the energy delivered to the circuit over which the integration extends, and the unit measured is usually the kilowatt-hour. This is the most frequently used revenue meter device on an electric utility system found in residences, commercial establishments, and industrial plants.

Wheeling charge. An amount paid by a consumer or utility for transporting power over electric lines owned by others. This can be in the form of an energy charge, demand charge, capacity charge, or a combination of these.

Index

A

D

E

F

M

N

S

T

U

V

W

Z